Advance Praise for *The Ecotechnic Future*

Greer's work is nothing short of brilliant. He has the multidisciplinary smarts to deeply understand our human dilemma as we stand on the verge of the inevitable collapse of industrialism. And he wields uncommon writing skills, making his diagnosis and prescription entertaining, illuminating, and practically informative. Not to be missed.

RICHARD HEINBERG
Senior Fellow, Post Carbon Institute
and author of *Peak Everything*

There is a great deal of conventional wisdom about our collective ecological crisis out there in books. The enormous virtue of John Michael Greer's work is that his wisdom is never conventional, but profound and imaginative. There's no one who makes me think harder, and *The Ecotechnic Future* pushes Greer's vision, and our thought processes in important directions.

SHARON ASTYK
farmer, blogger (SharonAstyk.com)
and author of *Depletion and Abundance*
and *A Nation of Farmers*

In *The Ecotechnic Future*, John Michael Greer dispels our fantasies of a tidy, controlled transition from industrial society to a post-industrial milieu. The process will be ragged and rugged and will not invariably constitute an evolutionary leap for the human species. It will, however, offer myriad opportunities to create a society that bolsters complex technology which at the same time, maintains a sustainable interaction with the ecosystem. Greer brilliantly inspires us to integrate the two in our thinking and to construct local communities which concretely exemplify this comprehensive vision.

CAROLYN BAKER
author of *Sacr*
Path
and

THE *ECO*TECHNIC *F*UTURE

THE *ECOTECHNIC FUTURE*

envisioning a post-peak world

JOHN MICHAEL GREER

First published in 2009 by New Society Publishers

This reprint published in 2024 by Sphinx Books, an imprint of Aeon Books

British Library Cataloguing in Publication Data

A C.I.P. for this book is available from the British Library
ISBN-13: 9781915952202

Cover Art by Phoebe Young

Contents

PART TWO – RESOURCES

Introduction

HISTORY HAS ITS turning points, the moments when pressures for change that have built over centuries finally burst through and reshape the world. Sometimes those moments can be sensed as they happen, but such moments of vision are rare; more often, the turning point can be traced only after the fact, when the failure of old patterns can no longer be ignored. In the gap between the coming of change and the moment of its recognition, old certainties fall to pieces and cannot be replaced, and plans for the future go awry because the future the planners have in mind fails to arrive.

A gap of this kind has become one of the most significant social facts of our time. The scale of that gap can be measured by the distance between the 21st century our society expected and the one that it got. Only a few decades ago, a galaxy of scientific pundits and media figures regaled an eager public with images of what the year 2000 would be like. There would be bases on the moon and a big wheel-shaped space station in orbit, with scheduled flights arriving there from Cape Canaveral twice a day. Undersea cities would dot the continental shelves and harvest the supposedly limitless resources of the sea. Clothing would be disposable and food would be synthesized on demand; fusion reactors would turn out limitless cheap power, geodesic domes would sprout everywhere and commuters would travel from lush suburbs to climate-controlled cities by helicopter instead of by car.

These fantasies were taken very seriously at the time — seriously enough to guide business decisions. In Seattle, where the 1962 World's Fair celebrated a 21st century full of space travel and triumphant technology as though it had already arrived, one forward-thinking builder in those years topped a new parking garage with a helipad and control tower, in hopes of getting a jump on the competition. As far as I know, no helicopter ever landed there, and the garage with its forlorn heliport was torn down during the housing boom of the late 1990s to make room for a block of singularly ugly condos. By that time, the rest of the future portrayed in the 1960s had suffered the same fate.

Behind this fact lay a simple but profound change in the foundations of the industrial world. Until the end of the 1960s, the biggest problem with energy resources in North America was how to keep the market from being drowned in too much oil. An obscure bureaucratic body, the Texas Railway Commission, had to restrict petroleum production in the United States to keep prices above production costs. Those days are gone, and not even the reckless overproduction that crashed the price of oil in the 1980s brought them back. The same sea change transformed the world's relationship to other natural resources, and to the natural cycles our civilization uses to absorb its wastes. In the 1980s and 1990s, the world's industrial nations used every economic, political and military lever they had to force down the prices of raw materials and the costs of pollution from their 1970s peaks, but the easy certainties of an earlier time had vanished.

Today the modern industrial economy seems as permanent as any human reality can be. That sense of permanence, though, is an illusion. The nonrenewable resources that went into building industrial civilization were vast, but they were never limitless. Their limits are now coming within sight, along with the equally strict limits of the Earth's ability to absorb the pollutants we dump into the skies, the seas and the land. In the aftermath of the 1960s, the industrial world could potentially have responded to the arrival of the limits of growth by launching a transition toward a sustainable future. For intensely human reasons, though, that was not done in time to make a difference.

The consequences of that failure to act have been examined in many recent books.[1] Their message is a sobering one: what remains of the Earth's natural resources and its capacity to absorb pollution will not allow us to continue living much longer the way we live today, much less provide for the endless progress nearly all today's political and economic ideologies expect. In place of the bright new world most of us anticipate, we are headed at a breakneck pace toward a future of narrowing options, dwindling resources and faltering technology, in which many of today's fondest dreams will face foreclosure.

This is the territory that my first book on the future, *The Long Descent*, set out to explore. That book studied the histories of other civilizations that had outrun their resource base, and used the parallels to map the trajectory our own future will most likely take: a ragged process of decline and fall unfolding over one to three centuries, ending in a dark age like the ones that followed the twilight of so many other civilizations. Many of the larger questions raised by that analysis, though, were left unanswered. It is one thing to recognize that today's industrial world is headed toward that difficult destiny. It is quite another to grasp what such a future implies, and to glimpse what the world will be like in the wake of our civilization's fall.

These more ambitious and quite possibly more foolhardy themes are the focus of this book. The reasons why the industrial age is ending and the immediate steps that can be taken by individuals and communities to cushion the descent are discussed in *The Long Descent* and many other books on the same topic.[2] Here, I will discuss these points only where they relate to the wider project of this book. Instead, *The Ecotechnic Future* will sketch the arc of history and human evolution in which the crisis of our time finds its context, and suggest actions that can be taken to make a better world not only for ourselves, but for our grandchildren's grandchildren.

The first section of this book, "Orientations," puts the decline and fall of the industrial age in the context of human ecology. Chapter One, "Beyond the Limits," traces out the historical arc of industrial society, explaining why the opportunity for a controlled transition to sustainability has already been missed, and what this implies for the future. Chapter Two, "The Way of Succession," shows how succession — a common ecological process — helps explain historical change, and shows that the end of the industrial age marks an early stage in a process leading in unexpected directions. Chapter Three, "A Short History of the Future," outlines the major forces already shaping the world our descendants will inherit, and Chapter Four, "Toward the Ecotechnic Age," suggests that succession may point toward the rise, centuries from now, of a new form of human civilization — an ecotechnic society — that will support a relatively complex technology while sustaining rich and sustainable relations with the rest of the biosphere.

The second section, "Resources," outlines the core elements of human society, and explores how individuals and communities can act now to help midwife a future ecotechnic age. Following a first chapter titled "Preparations," which explores potentials for constructive action and challenges some common assumptions about the future, seven thematic chapters — "Food," "Home," "Work," "Energy," "Community," "Culture," and "Science" — examine the opportunities offered by these basic elements of human life, and build a case for pursuing diversity and experimentation as central strategies for the long transition to the ecotechnic age. The third and final section, "Possibilities," contains a single chapter, "The Ecotechnic Promise," that places the emergence of the ecotechnic age in the context of human history as a whole, and explores some of the questions of meaning that give history its importance.

As always, I have benefited from the help of many minds in writing this book. Many of the ideas presented here were first aired in essays posted on my blog, "The Archdruid Report," and the comments and criticisms received from readers of the blog have had a crucial role in shaping the argument of this book; I owe thanks to all those who participated. I would also like to thank Sharon Astyk, Rob Hopkins, Bill Kauth and Stuart Staniford for dialogues that helped me refine my ideas. In the contemporary Druid community, my spiritual home and the foundation of many of my ideas, I owe particular thanks to Philip Carr-Gomm, Gordon Cooper, Siani Overstreet, Tully Reill and all the members of the public e-mail forum of the Ancient Order of Druids in America (AODA), the Druid order I head. Last to be named, but first in her influence on my thought and my life, is my wife Sara. My gratitude goes out to all.

PART

I

ORIENTATIONS

Beyond the Limits

THE CRISIS OF industrial civilization that dominates so many of today's headlines has been explored from a dizzying range of perspectives. Political scientists and economists have talked about the ways that contemporary governments have backed themselves into a variety of corners, and how market systems have run off the rails. Other social sciences have had their contributions to make, and so have less scientific fields of study such as history, philosophy and religion. Still, the roots of today's crisis reach down into a disastrous mismatch between today's human societies and the world of living nature on which human life depends.

This realm of root causes can best be summed up in that much-abused word "ecology." A few moments to clarify what the word actually means, as distinct from the political and cultural baggage that has been heaped on it, are thus probably worth spending. Ecology comes from the Greek words *oikos*, "home" and *logos*, "speech." Ecology, then, is "speaking about the home"; less poetically, it is the scientific study of the relationships between living beings and their environments. It embraces all living things, including the ones who are able to read this book; human ecology applies the principles learned from studying other living beings to the relationships that connect human beings with their environment.

Central to ecological thought is an awareness of connections. No living creature, nor any species of living creatures, exists by and for itself; rather, everything living and nonliving in a given area is part of an integral whole that ecologists call an ecosystem. Many links weave each part of an ecosystem together with the others, but several kinds of connection play leading roles in most ecosystems. Of these, the most important are defined by the flows of energy that enable life to exist at all.[1]

These flows take different forms as they pass from one living thing to another. To the grasses in a meadow, for example, the energy that matters is the sunlight that falls on their leaves and powers the everyday miracle of photosynthesis, turning carbon dioxide and water into the sugars that grasses need to survive. For the field mice that live in the same meadow, the energy source that matters is the grass. Since the grass gets its energy from the sun, the field mice, in turn, live on secondhand sunlight. The fox that eats the mice and the fungi that grow on mouse droppings get their own share of sunlight at another remove. Track the energy from sun to grass to mice to fox and fungi, following what ecologists call a food web, and you learn a great deal about the ecosystem in which they're embedded.

Minerals, water, oxygen and certain other substances also move through ecosystems and support life. There's an important difference, though, between matter and energy. Energy always moves from higher to lower concentrations until it is dispersed and lost as waste heat; all energy flows are thus one-way streets. Matter, though, moves in circles whenever circumstances permit. Grasses in a meadow need phosphorus, and get it from the soil; mice need phosphorus, and get it from grass; fungi need phosphorus, and get it from mouse droppings; and from the fungi it passes back into the soil, to be taken up once again by the grass. In most healthy ecosystems, very few nutrients leak out of these cycles, and what does escape is quickly taken up by other ecosystems nearby, while energy collected by the system in times of relative abundance is stored in chemical form for times of relative scarcity. The result is a stable system.

Not all ecosystems can manage this kind of stability, though. Ecosystems in small forest ponds, for example, often get their energy and minerals from falling leaves.[2] If the leaves fall steadily, as they do in most tropical forests, the pool ecosystem can reach stability, with energy flowing at much the same rate from season to season and minerals cycling from one organism to another. If the leaves fall all at once, though, as they do in many temperate forests, most of the energy the pond gets in a year arrives in a single pulse. Ecologists call the result "overshoot": the pond dwellers feed freely until the food runs out, and then most or all of them die.

The same thing can happen when variations in energy supply are less dramatic. The energy used by field mice in a meadow, for example, depends on variations in climate that affect the grass crop. When the climate is favorable, there may be plenty of grass and the mouse population increases. When the climate turns harsh and the grass is stunted, starvation reduces the mouse population. If several good years are followed by a bad year, the mouse population can climb well above carrying capacity — the number of mice per acre that the meadow can support over the long term — only to drop well below carrying capacity as starvation and disease take their toll.

On the far end of the overshoot spectrum are situations in which a community of living things receives a surplus of energy through some accidental event that happens only once. Imagine how the lives of field mice would be transformed if a truck full of grain overturned on the nearby freeway and spilled its load in their meadow. The mice suddenly have more energy than they can use and their population soars far beyond the meadow's carrying capacity. As the number of mice in the meadow grows, though, the rate at which the grain is consumed also rises, until the grain begins to run short. At this point nothing the mice can do will spare them from dieoff; most of the mice will starve, and the survivors' struggle to keep themselves fed may damage the meadow badly enough to decrease its carrying capacity over the long term. Years later, the meadow may still not support as many mice as it did the day before the truck overturned.

A population of living beings doesn't have to exhaust all the resources of its environment to go into overshoot. It only takes the depletion of one vital resource to cause population growth to give way to dieoff. This principle of ecology is known as Liebig's law of the minimum. Mice in a meadow have relatively simple needs — food, water, and shelter from weather and predators — but a shortage of any one of them can keep mice scarce or absent in an otherwise inviting habitat. Life forms with more complex needs have many more points of vulnerability, and just as with the mice, a protracted shortage of any one necessity can make survival difficult or impossible.

human ecologies

The doings of mice in a meadow may not seem to have much in common with the decline and fall of industrial civilization, but there's a direct link. Human societies, like other ecosystems, can be understood by tracing the flows of energy and cycles of matter that keep them going. In the oldest form of human ecology, the hunter-gatherer system that nurtured most of the human cultures that have ever existed, the energy paths that keep human beings alive pass through wild nature and take shape as the plant and animal foodstuffs that human beings eat, with a little extra in the form of firewood from trees. Most substances in a hunter-gatherer ecology move through cycles not that different from the ones in a meadow full of mice, but not all: a few materials — stone, bone, wood and the like — enter the human ecology as raw materials for tools and crafts, rather than as minerals and nutrients in food, and define a new kind of ecological relationship. Problems with that relationship, in turn, form the root cause of the crisis of the industrial world.

In nonhuman nature, as we've seen, energy follows one-way paths, while matter moves in cycles whenever conditions permit. Most of the raw materials used for tools and crafts in human cultures, by contrast, follow the same sort of one-way paths as energy. A piece of flint can be taken from the ground and crafted into a hand axe by a hunter-gatherer; ordinary wear and tear chips flakes from it, and more have to be chipped off periodically to renew the edge; finally the piece that's left is too small to use, and gets thrown aside to be found by some future archeologist. Nearly all of the materials used in human culture move in these straight lines, rather than in self-renewing cycles.

Not all materials are equal, though. Some replace themselves quickly, others slowly, and still others will not renew themselves in the lifetime of the Earth. These differences have massive impacts on the history of the human cultures that depend on them. All other things being equal, a human ecology that relies on fast-renewing resources such as annual plants tends to be more stable than one that depends on slow-renewing resources such as wood, which yields steady supplies only so long as the forests are not harvested too greedily. When a culture depends on a resource that is nonrenewable within human time frames, the one-way trajectory of cultural materials can turn into a lethal threat: any use speeds the arrival of the moment when useful amounts of the resource no longer exist.

All these dynamics apply equally to simple and complex human ecologies. The village farming ecologies that spread around the world after the last ice age, the nomadic herding ecologies that came into being a few millennia later, and the urban-agrarian ecologies that first rose in the Middle East around 6000 BCE all depended on a mix of resources with different renewal rates, including some that were nonrenewable. The nature of the mix had a potent impact on the fate of each society; those that relied primarily on resources that renew quickly, or paced their use of those with slower renewal rates, turned out to be far more durable than those that relied mostly on nonrenewable resources, or used renewable ones faster than nature could replace them.

The industrial system, the dominant human ecology on Earth just now, is more dependent on nonrenewable resources than any other, and the resources it uses differ from those of other human ecologies in another important way. Always in the past, the food that powered human and animal muscles, along with modest amounts of sun, wind, water and biomass, provided nearly all the energy used by human societies. All these resources were at least potentially renewable, because they drew their energy from the Earth's current solar input — the flood of diffuse but steady energy that streams down from the sun every day, driving the weather systems that power the winds and recycle water into rain and snow, converting carbon dioxide and water into sugars in the leaves of green plants and providing direct warmth that human beings have used in countless ways. Human societies could and did come to grief by damaging the soils, forests, watersheds and wetlands needed to turn sunlight into useful energy, but the flow of solar energy itself was renewed with every dawn.

Industrial civilization broke that mold by turning from current solar input to ancient sunlight stored up as fossil carbon during the prehistoric past. This allowed industrial nations to use more energy than any past society, but that advantage came with a hidden price tag. By making industrial civilization dependent on a nonrenewable resource, it placed humankind in the position of the mice in the meadow after the grain truck overturned.

The staggering scale of the modern world's dependence on fossil fuels has to be grasped to make sense of the resulting predicament. The world currently uses around 84 million barrels of oil, 12.5 million short tons of coal, and 8 billion cubic meters of natural gas every single day.[3] This torrent yields as much energy as one-fifth the sunlight absorbed by all the world's green plants each day[4] — an astonishing amount of energy for a single species. Nor is this energy evenly distributed; most of it is used by a handful of the world's nations, with the United States alone accounting for a quarter of the total. Almost everything that sets today's industrial humanity apart from nonindustrial cultures is a product of the intensive exploitation of petroleum and other fossil fuels. Most people who live in the sprawling, energy-intensive cities of North America, Europe and Japan, however, take it for granted that the fossil fuel energy that fills their lives is normal and permanent, and a great many of them assume that someday all of the world's people will live in similar surroundings.

It's unlikely that the field mice in a meadow full of grain from a spilled truck, or the little lives in a forest pond temporarily brimful of fallen leaves, think in these terms. Still, their situations and that of today's industrial societies are too close for comfort. Every barrel of oil, ton of coal or cubic meter of natural gas we burn today is gone forever, and after three hundred years of breakneck extraction, the limits are coming into sight. As well as energy, most of the materials that sustain the industrial way of life are also nonrenewable or are being used faster than nature can replace them. Like the man in the proverb who sawed off the branch he sat on, industrial civilization is burning through the resources that make its existence possible, and even the less industrialized societies have become so dependent on the industrial system that their survival too is in doubt as fossil fuels run short.

Petroleum, the most energy-rich and economically significant of the fossil fuels, is also the most depleted, and the consequences are already beginning to affect the world's economic and political systems. Talk about "peak oil" — the point at which roughly half the world's conventional petroleum reserves have been pumped out of the ground and production worldwide begins to decline — is finding its way into the media and popular culture. Most official estimates place the arrival of peak oil sometime between 2020 and 2030, close enough that efforts to deal with the resulting energy crisis ought already to be under way, but these estimates have already been overrun by events. World production of conventional petroleum, in fact, peaked in 2005 and has been in a shallow decline ever since.[5]

That decline has been masked so far by rising production of natural gas liquids, tar sand extractives and biofuels, but these are not really replacements for conventional petroleum. Natural gas liquids are subject to sharp depletion problems of their own, while the others all require large investments of energy, most of which comes from petroleum. In a real sense, to count ethanol production as part of petroleum production without subtracting the diesel fuel, petroleum-based agricultural chemicals and other inputs needed to grow the ethanol feedstock and convert it into ethanol is to count the same oil twice over. The same issue holds for the massive energy inputs needed to turn tar sand into barrels of oil and for other nonconventional fuels as well. For the time being, this dubious number juggling has kept production figures up, which may well be the point of the exercise.

Yet the economic consequences of the peak of petroleum production are already beginning to show themselves. During 2007 and early 2008, fuel and chemical prices soared to previously unimaginable levels and the nonindustrial world faced desperate shortages of petroleum products. Meanwhile the diversion of grains into ethanol production sparked dramatic price hikes and food shortages around the world. The economic crisis of the second half of 2008 sent prices plunging again as speculators squeezed by an imploding real estate market sold off their energy investments. Still, as I write these words, oil sells for around $45 a barrel — a price that would have been considered very high a few years ago — and production is dropping as oil wells and alternative sources that are only economical above $100 a barrel are shutting down. All the ingredients of a new price spike are in place.

Economists have long insisted that rising prices for any commodity would automatically bring about increased supply, but oil has broken this supposed law with impunity: worldwide, oil production has been flat for most of a decade while prices soared, and plenty of energy sources that were touted as ready for the market once oil broke $30, or $40 or $50 a barrel are still nowhere in sight. Thus the world has reached the point at which geology trumps market forces, and supply can no longer increase to meet the potential demand. Speculation and a disintegrating economy have turned the rising curve of energy prices into a volatile landscape of price spikes and crashes.

The resulting predicament can be stated simply enough. The industrial world no longer has the resources or time to change fast enough to stave off its own decline and fall. All energy sources are fully committed to existing needs, and any attempt to free up resources for some new project will conflict with the demands of existing economic sectors. The US government may be in a position to loan Wall Street $700 billion it doesn't have — in today's economic world, money is so close to a mass hallucination that it's not surprising to see it wished into being so casually — but actual resources such as fossil fuels, trained labor forces, and time are not so flexible. In the shadow of these unmentionable realities, the world is hurtling toward an unwelcome future for which most of us are hopelessly unprepared.

tomorrow comes anyway

The irony in this situation is that the industrial world started to get ready for the end of the petroleum age in the 1970s, but abandoned its preparations thereafter and moved in the other direction. It's worth revisiting that time of lost opportunities, not least because most of today's debates about energy and the future are simply rehashes of equivalent debates that filled the pages of specialist journals and mainstream publications thirty years ago.[6] The energy crises of that decade played a large role in sparking those debates, but another factor was at least as important: the emergence of ecology as a major theme in the collective discourse of the time.

In the 1960s, the impact of pollution on the industrial world's air, water and land reached levels that could no longer be ignored. The rise of environmentalism as a political force was fiercely opposed by political, business and scientific establishments; when Rachel Carson's epochal *Silent Spring* appeared in 1964, it was denounced as crackpot pseudoscience not only by the chemical industry but also by some of the nation's most prominent scientists. Still, when rivers in industrial districts were so full of industrial wastes that they routinely caught fire on hot summer days — as happened in several parts of North America in the 1960s and 1970s — business as usual became hard to defend.

The new environmentalism of those years overturned, at least for a time, some of the basic assumptions of the status quo. Energy issues were only part of a broad spectrum of innovation during those years, which saw major breakthroughs in many fields. Still, a great deal of work went into dealing with the twilight of fossil fuels and the need to make the transition to renewable energy. By the end of the decade, a consensus had begun to form in the energy field that drastic increases in efficiency and conservation were possible as well as necessary, so that the modest energy supplies available from sun, wind, water and other renewable sources could meet the reduced needs of a future "conserver society."

Unfortunately, human societies do not always choose reasonable options. As the 1980s began, conservative politicians throughout the industrial world found that they could drive their liberal opponents from power by insisting, among other things, that nobody had to make sacrifices to achieve sustainability because the market would take care of the energy problem. Two key advantages allowed them to push this shortsighted but successful strategy. First, soaring energy costs in the previous decade drove conservation campaigns that cut energy use worldwide by a significant amount — petroleum consumption, for example, dropped by 15 percent between 1975 and 1985 — and this put significant downward pressure on energy costs. Second, reckless pumping of Alaska's North Slope and the North Sea fields allowed the United States and Britain to flood an already well-supplied market with oil.

The resulting price crash dropped the price of petroleum to US$10 a barrel — corrected for inflation, the lowest price in history — and kept it down for most of the quarter century from 1980 to 2005. During that time, petroleum could be had throughout the world at prices so low that they put no economic burden on anyone but the very poor. Other energy costs slumped accordingly, as cheap oil competed with other fuels for market share and also provided a hidden cost subsidy to other fuels. The cost of mining coal, for example, went down as the price of the diesel fuel that powers coal-mining machinery dropped.

Those energy sources that could not compete with ultracheap fossil fuel, even with the energy subsidy from oil, became obsolete. Renewable energy industries in the United States and Europe suffered drastic contractions; the nuclear industry, which was already struggling, took even heavier economic hits and survived only through massive government subsidies. After 1980, the thought of a transition away from fossil fuels vanished from conventional wisdom, as an economy based on extravagant energy use took shape around the globe.

Because transportation costs were negligible, manufacturers could freely arbitrage labor costs between industrial and nonindustrial countries. The resulting flood of cheap consumer products created a throwaway economy in which anything that stopped working, or even stopped being popular, went to a landfill. Fuel-efficient cars were replaced by huge SUVs and the Hummer, the military assault vehicle turned suburban land yacht that became the definitive symbol of an age of excess. Meanwhile, the fossil fuels that might have powered the transition to a sustainable future were wasted on a quarter century of extravagant living.

The consequences of this immense mistake are only just beginning to be felt today. It was clear by the 1970s that the transition to a new energy economy had to happen while there was still a surplus of fossil fuels. Influential studies stressed that by the time actual shortages began, all existing resources would already be committed to meet existing needs. Trying to allocate them for other uses would cause drastic shortfalls in sectors of the economy that were using those resources already.[7] Once the surpluses were gone, the industrial world would no longer be able to power the present and prepare for the future at the same time. Eventually, as energy supplies depleted further, what remained would no longer do either one.

These concerns, it turned out, are not merely theoretical: the recent ethanol boom shows the process at work. By shifting grain supplies from the human food web to ethanol production, gasoline supplies were boosted over the short term, but this also sent food prices soaring, sparking inflation across a wide range of products, causing a cascade of problems elsewhere in the economy and sparking food riots in poor countries around the world. Imagine the results of a similar attempt when fossil fuel supplies are dwindling and the energy and raw materials needed to build ethanol plants have to be taken away from some other sector of the economy; the potential for serious trouble is hard to ignore.

In hindsight, the entire period from 1980 to 2005 will likely be seen as one of history's supreme blind alleys. For a quarter century or so, people across the industrial world consumed energy as though there was no tomorrow. The problem with that way of living, of course, is that tomorrow comes anyway. The economic convulsions and energy shortages shaking the world today are serving notice of that unwelcome reality. The lesson these troubles bring home is that the economic arrangements, the infrastructure and the personal and collective habits that grew up in response to a misplaced faith in perpetual abundance make no sense in a world subject to ecological limits.

Still, these facts are nowhere to be found in our collective discourse just now. The stores of fossil carbon locked up inside the Earth in prehistoric times are dwindling fast, and no other energy source as highly concentrated and easily accessible is to be found on our planet. We cannot continue living a lifestyle dependent on the highly concentrated, cheap and abundant energy only fossil fuels can provide, and we have already used up the energy surpluses that might have made it possible for us to make a controlled transition to some other way of life. We no longer have the option of avoiding crisis; the only questions that remain are how to deal with the consequences. Yet discussions about the future nearly always start from the unstated assumption that there are still ample supplies of energy, resources and time to deal with any crisis or overcome any challenge.

Consider how this attitude pervades current debates about energy. One side claims that we simply need to find another energy source as cheap, abundant and concentrated as petroleum, and run our society on that instead. The other claims that we need to replace every part of our society that depends on cheap, abundant, concentrated energy with others lacking that dependence and run this radically reconfigured society with alternative resources instead.

The problem with the first belief is simply that there is no other energy source available to us that is as cheap, abundant and concentrated as petroleum, or equivalent to any of the other fossil fuels. The second belief faces the same problem as well as another of the same order: if we toss aside the parts of our society that depend on cheap, abundant, concentrated energy, nothing remains.

An entire world of infrastructure and technology would have to be replaced in a hurry, and the components for a new low-energy society are not simply sitting on a shelf somewhere, waiting to be used. Meanwhile industrial civilization goes on consuming irreplaceable resources at alarming rates and any slowdown in consumption leads government and business interests alike to cry for immediate remedies to speed the economy back up. It is hard to think of a better recipe for a difficult future.

the illusion of independence

Behind the bizarre spectacle of a civilization sleepwalking toward the abyss lies the failure of nearly all sides in today's debates to grasp the most basic elements of ecological reality. To an astonishing degree, defenders and challengers of the status quo both think and act as though human society need not take its relationship to nature into account when making important decisions. At most, some narrowly defined issue — problems with the ozone layer, say, or the awkward fact that oil Americans want to burn turns up under someone else's real estate — gets a flurry of attention when it can no longer be ignored. The hard reality that every political, economic and social decision made by a modern society needs to take impacts on nature into account has been absent from our collective dialogue for a quarter century.

This blindness to the basis for human survival makes no sense in even the most pragmatic terms. In 1997 a team of ecologists and economists headed by Robert Costanza published the results of a study estimating the economic value of nature.[8] Such calculations inevitably miss more than they grasp, but the goal was sound; the team hoped to determine if it really made economic sense to do as human societies have done for three hundred years, treating the natural world as something valueless except as a source of raw materials and a place to dump waste. They worked out what it would cost to replace the services that the Earth provides free of charge to its human inhabitants. Those services turned out to equal, on an annual basis, around three times the value of all human economic activity. If the global economy had to pay for the services it gets from nature, in other words, it would be bankrupted three times over in a single year.

Our dependence on nature can also be measured in much less abstract terms. Consider the oxygen we breathe. It doesn't just show up by accident; it's put there by green plants. If the plants go away, so does the oxygen, and so do we. The Earth's supply of fresh water, similarly, is renewed by atmospheric cycles vulnerable to human mishandling. The experiment of producing food by dumping petrochemicals into nearly sterile soil — which is how modern chemical agriculture works in practice — is proving to be a failure in the long term; as the industrial age ends, the natural cycles that support sustainable agriculture are the only option remaining to put food on humanity's table and keep us all from starvation.

The contemporary belief that science and technology have freed humanity from dependence on nature is thus a dangerous illusion. It's this illusion that leads so many well-intentioned people to argue that nature is an amenity, and should be preserved because, basically, it's cute. That sort of argument invites the response, just as stereotyped and more appealing to our culture's habits of thought, that hard-headed practicality takes precedence over emotional appeals and nature can therefore be ravaged with impunity.

Yet nature is not an amenity, and the "practicality" that leads people to ignore ecological realities typifies the sort of thinking C. Wright Mills called "crackpot realism,"[9] the use of rational means to pursue hopelessly irrational ends. If anything, industrial civilization is the amenity, and it's not particularly cute. Nature and humanity can survive without industrial civilization, but neither industrial civilization nor humanity can survive without nature — no matter how hard we pretend otherwise, or how enthusiastically we stuff our brains with fantasies about electronic reincarnation and the good life in deep space.[10]

We have all grown up, one might say, thinking of nature as an adorable, helpless bunny that some people want to protect and others, motivated by the will to power that is the unmentionable force behind so much of contemporary culture, want to stomp into a bloody pulp just to show that they can. Both sides are mistaken, for what they have misidentified as a bunny is one paw of a sleeping grizzly bear who, if roused, is quite capable of tearing both sides limb from limb and feasting on their carcasses. The bear, it must be remembered, is bigger than we are, and stronger. We forget this at our desperate peril.

A crucial fact frames this analysis. Despite its strident insistence on its own uniqueness, industrial civilization is simply one example of a common trajectory. Many other civilizations before ours have overshot the limits of their resource base, and stumbled down the ragged curve of decline and fall.[11] Our civilization became more gargantuan than others through the luck and cleverness that enabled it to replace sun, wind, water and muscle with the vast but finite supplies of ancient sunlight stored away in the Earth's fossil fuels.

Still, the course of decline and fall traces the same trajectory across many different geographical scales. Local civilizations restricted to a single bioregion and limited to stone tools, such as the ancient Maya, rose and fell in much the same way as sprawling empires built on a continental scale and equipped with sophisticated technology, such as ancient Rome. It does not require a leap of faith to suggest that the same patterns will also shape the fall of an industrial civilization that includes several continents and has dominated, for a brief period, the rest of the planet.

The insights of ecology, however, offer a wider perspective on the changes ahead of us. By the light of these insights, the crisis of our time takes on an unexpected shape, and can be understood as an unrecognized manifestation of a common ecological pattern. That perspective puts paid both to the fantasy that industrial civilization stands at the summit of history and to the fear that its decline and fall marks history's end. Rather, the road to the future leads along unexpected paths toward a phase of human history I have named the ecotechnic age.

The Way of Succession

COMPARISONS THAT PUT human behavior on a par with field mice in a meadow have become unpopular in recent years. They typically bring the retort that human beings are smarter than other life forms and the parallel therefore doesn't apply. Flattering to human vanity as these objections may be, a certain skepticism comes to mind in the light of current events. For most of a decade now, petroleum production has been stuck in a narrow band despite soaring prices and massive drilling programs. Many countries in the nonindustrialized world are already in desperate straits, forced to ration electricity and fuel, as their access to fuel dries up. Meanwhile government and business leaders in the world's industrial nations, which have even more to lose from the twilight of cheap abundant energy than their poorer neighbors, are still treating the twilight of the age of oil as a public relations problem. If we're enough smarter than field mice to escape their fate, we have yet to demonstrate it.

On a deeper level, comments such as this last miss the point just as thoroughly as the claim they're meant to satirize. The value of the field mouse metaphor is that it shows how ecological processes work in a context simple enough to permit clarity. The same patterns can be traced in more complex systems, human societies among them. The pattern that pushes field mice into overshoot, after all, is the same pattern that has driven collapse in many human cultures of the past. If a culture uses any one of the resources necessary to its survival at an unsustainable rate, it follows the usual overshoot curve — population boom, followed by population crash.[1]

A classic example of the process took place on Easter Island, where towering stone statues still serve as a reminder that cultural evolution is not a one-way street.[2] The people of Easter Island thrived on the rich sea life of the Pacific, but their access to those resources depended on wood for canoes, and that turned out to be their downfall. It took the Easter Islanders only a few centuries to strip the forests of their island home to bare soil. Once deforestation was complete, they were left with a population far beyond what the island's own resources could support. The result was a bitter spiral of starvation and violence that brought a thriving culture to its knees. Note the role of Liebig's law of the minimum: the loss of one essential resource made an entire human ecology unworkable.

Thus humanity is no more exempt from ecological processes than it is from the law of gravity. This comparison is worth taking further, as the invention of airplanes doesn't mean that gravity no longer applies to us; it means that if we use plenty of energy, we can overcome gravity and leave the ground for a little while. In much the same way, extravagant and unsustainable energy use lifted a minority of the world's population high above subsistence level for a while, but that doesn't mean that ecological laws no longer affect us. It means that for three hundred years, we've been able to ignore the Earth's carrying capacity by burning huge amounts of fossil fuels. When the fossil fuels are gone, the laws and limits will still be there.

Recognizing the ecological dimensions of human life has an unexpected benefit: some of the patterns at work in humanity's relationship with nature take on a new meaning. What appears to us as collapse, seen from an ecological perspective, becomes a familiar environmental pattern: the process of succession. The fact that this isn't a household word across the industrial world may be one of the best signs of our isolation from the realities of life on this planet.

Fortunately, succession is an easy process to grasp. Any of this book's readers who were unwise enough to buy a home in one of the huge and mostly unsold housing developments cranked out at the top of the late real estate bubble will be learning quite a bit about succession over the next few years, so summarizing it here may be useful for more than one reason.

succession in action

Imagine, then, an area of bare bulldozed soil left behind by some bankrupt developer, in some corner of the country where the annual rainfall is high enough to support woodland. Long before the forlorn sign saying "Coming Soon Luxury Homes Only $650K" topples to the ground, seeds blown in by the wind will send up a first crop of invasive weeds. Those will ready the ground for other weeds and grasses, which will eventually choke out the firstcomers. After a few years, shrubs and pioneer trees will sprout, becoming anchor species for a young woodland, which will shade out the last of the weeds and the grass. In the shade of the pioneer trees, saplings of other species will find homes. If nothing interferes with the process, the abandoned lot will pass through as many as a dozen distinct stages before it finally settles down as an old growth forest a few centuries in the future.

This is what ecologists call succession.[3] Each step along the way from bare dirt to mature forest is called a *sere* or a *seral stage*. The seres can be tracked just as precisely in the animal population of the vacant lot: one species after another moves into the area for a time, until it's supplanted by another species better adapted to the changing environment and food supply. Succession also proceeds underground, as the rich fabric of life that makes up healthy soil reestablishes itself and cycles through its own changes. Watch a vacant lot in a different ecosystem with, for example, much less rainfall, and you'll see it go through its own sequence of seres, ending in its own climax community — the term for the final, relatively stable sere in a mature ecosystem, like the old growth forest in our example.

Compare the process of succession to the population cycles of field mice in a meadow discussed in Chapter One and an important distinction surfaces. The mice, their food supply and the entire dance of energy flows and nutrient cycles that allows them to survive in the meadow exist on one time scale; succession operates on a much larger one, and generations of mice can live and die while the ecosystem they inhabit passes through a single sere. When that sere gives way to another and the available niches change, the field mice themselves will leave, and another species such as wood mice will take their place. The same ecological laws govern the field mouse and wood mouse seres, but changed conditions produce different ecosystems.

The shift from sere to sere is driven by differences in the way that the species in each sere deal with energy and materials. Species in early seres — R-selected species, in ecologists' jargon — maximize their control over resources and their production of biomass, even if this requires inefficient use of resources. Weeds are classic R-selected species: they grow fast, spread rapidly and get choked out once slower-growing plants get established, or the abundant resources that make their fast growth possible run short. Species of later seres — ecologists call them K-selected species — maximize their efficiency in using resources, even when this means accepting limits on biomass production and expansion. Temperate zone hardwood trees are classic K-selected species: they grow slowly, take years to reach maturity and endure for centuries if undisturbed.

The division between R-selected and K-selected species has a close relationship to the process of overshoot discussed in Chapter One. For an R-selected species, overshoot is business as usual; their seres often last only a very short time, and their strategy for survival hinges on reproducing abundantly enough that some offspring will get through the inevitable crash. For a K-selected species, on the other hand, overshoot happens only when some outside force disrupts the normal cycles of their environment; their survival strategy hinges on stability. Generalist species like field mice can tip one way or the other, depending on environmental factors; if food supplies are stable, they will follow a K-selected strategy, while the more unstable the food supply, the more R-selected they will become.

Human beings are among Earth's most successful generalist

species, and the flexibility of human culture means that selection takes place on the level of behavior more often than that of genetics. The same evolutionary patterns, however, still apply: human communities compete with one another for resources, and the differing behaviors they foster in their members include close analogues to the R-selected and K-selected strategies found among other species. Most of humanity today is in the early stages of a transition between an extremely R-selected sere and the more K-selected sere that will replace it. The industrial economies of the present, like any other R-selected sere, maximize production at the expense of sustainability; like weeds, they spread fast, use resources recklessly, overshoot the carrying capacity of their environment and perish. Like any other K-selected sere, the successful human ecologies of the future, in a world without today's cheap abundant energy, will need to maximize sustainability instead.

To grasp the whole picture, though, it's necessary to shift to a third time scale, because climax communities are stable only over the time span of centuries; beyond that, new factors come into play. The climax forests that European settlers found in North America, for example, had not been there forever; products of a complex balance of climate, resources and evolution, they had been going through changes of their own since the end of the last Ice Age. Environmental shifts change climax communities, and so does the arrival of new species through migration or evolution. Sometimes these newcomers make succession move in reverse for a while, when an invasive R-selected species outcompetes dominant K-selected species in a climax community. After the new sere settles into place, the succession process starts moving again, but the new climax community may not look much like the old one.

Apply this model to the human ecology of North America and fascinating patterns emerge. After 1492, climax communities of K-selected Native American farmers and hunter-gatherers were displaced by an invasive, R-selected population of European farmers. Shortly after the invasive community established itself, and before succession could push it far in the direction of a more K-selected ecology, a second invasive sere — the industrial economy — spread rapidly into the same ecosystem. This second invasive sere, the first of its kind on the planet, was on the far end of the R-selected spectrum; its ability to access extravagant amounts of energy enabled it to dominate the farming sere that preceded it, and push what was left of the old native climax communities to the brink of extinction.

Like all R-selected seres, though, the industrial economy was vulnerable on two fronts. First, it faced the certainty that a more efficient K-selected sere would eventually outcompete it. Second, its need to use resources at unsustainable rates made it vulnerable to cycles of overshoot and crash that would eventually give a more efficient sere its opening wedge. Both those events are now under way. At this point, the industrial economy is well into overshoot and a crash of some kind is inevitable. Meanwhile, the more K-selected human ecologies of the next sere have been sending up visible shoots since the 1970s, in the form of small organic farms, local farmer's markets, appropriate technology and alternative ways of thinking about the world.

Three points need to be made in this context. First, as mentioned above, human beings can adapt to new conditions by changing their culture and behavior; they are not limited by the slower pace of genetic drift and selection. Thus the shift from an R-selected to a K-selected human ecology need not involve mass dieoff. At this point, it's unlikely that a K-selected human ecology will be expanded fast enough to take up the slack of the disintegrating R-selected industrial system, and so there's likely to be a great deal of human suffering and disruption over the next few centuries. Still, those individuals willing to transition to a K-selected lifestyle may find that the disintegration of the industrial system opens up opportunities to survive and even flourish.

Second, the concept of succession suggests a radically different view of what an advanced civilization might look like. Modern industrial society is the exact equivalent of the first sere of pioneer weeds on a vacant lot — fast-growing, resource-hungry, inefficient and at constant risk of replacement by more efficient K-selected seres as the process of succession unfolds. A truly advanced civilization might well have more in common with a climax community: it might use very modest amounts of energy and resources with high efficiency, maximize sustainability and build for the long term.

Third, the climax community that emerges after a prolonged ecological disruption and the arrival of new species rarely has much in common with the climax community before the disruptions began. In the same way, and for many of the same reasons, claims that the deindustrial world will end up as an exact equivalent of some past society — be that medieval feudalism, tribal hunter-gatherers or anything else — need to be taken with more than the usual skepticism. Rather, the climax community toward which the successional process of modern history is heading may be something that has never before existed on this planet.

succession and agriculture

Agriculture, the basis for human survival in nearly all the world's societies just now, offers a sharp lesson in the way succession works. It's common these days for people to assume that today's industrial agriculture is "more advanced" and therefore better than the alternatives. It's certainly true that the industrial approach to agriculture outcompeted its rivals in the market economies of the 20th century. Fossil-fueled machines had the advantage over human and animal labor, and fossil fuel-derived chemicals over natural nutrient cycles, because the energy and raw materials provided by fossil fuels were so much cheaper and more abundant than the alternatives. Still, in any ecological sense, modern industrial agriculture is *less* efficient than the alternatives.

Another example of succession from nature will help clarify the distinction. In the upland pine forests of central Oregon, fireweed — a pioneer plant, and strongly R-selected — grows in the aftermath of forest fires, thriving on the abundant minerals concentrated in wood ash and on bare disturbed ground where it can monopolize whatever nutrients the soil offers. As it spreads, though, it uses up the rich nutrient concentrations that allow it to thrive; the nutrients follow a linear path rather than a cycle, and so the plant leaves the soil less suited to its own survival. Finally, other plants better adapted to less concentrated nutrients replace it.

By contrast, the climax community in those same central Oregon drylands is dominated by pines of several species, and in a mature pine forest most nutrients are either in the living trees themselves or in the thick duff of fallen pine needles that covers the forest floor. The duff soaks up rainwater like a sponge, keeping the soil moist and preventing nutrient loss through runoff; as the duff rots, it releases nutrients into the soil where the pine roots can access them, and also encourages the growth of symbiotic soil fungi that improve the pine's ability to access nutrients. Thus the pine creates and maintains conditions that foster its own survival.

The same distinction can be traced just as clearly in the history of agriculture.[4] The first systems of agriculture emerged in the Middle East sometime before 8000 BCE, in the aftermath of the drastic global warming that followed the end of the last ice age and caused ecological disruption throughout the temperate zone. In the Middle East, grasslands turned into desert as winter rains that had fallen reliably for millennia stopped, and local tribes that had subsisted by hunting and gathering for millennia turned to grain cultivation in river valleys and livestock raising on the hills as the only alternative to starvation. The same process took place somewhat later in Mexico, the heartland of New World agriculture, as a parallel set of climate shifts caused desertification there as well.

The new human ecology of farming proved highly successful and spread rapidly, but in its first forms it was highly inefficient; early agriculture was an R-selected system, relying on natural soil fertility and depleting it rapidly in the process. The result was repeated disaster, as early farming societies exhausted their soil and collapsed. It took thousands of years and a great deal of trial and error before more K-selected agricultural systems replaced the original models, and some of the final steps in that process of succession did not take place until the birth of organic agriculture in the 20th century. Still, it's important to realize that it did become sustainable. The immense stability of Asian rice culture was documented long ago by F.H. King in *Farmers of Forty Centuries*; few people realize, however, that Syria — where grain farming was probably invented, and has certainly been practiced as long as anywhere on Earth — is still a major wheat exporter today.

Thus the first inefficient methods of growing grain were equivalents of pioneer weed seres. Early agriculturalists used nutrients inefficiently, depleting topsoil in the process. This guaranteed that eventually they would become their own nemesis. The more sustainable methods of agriculture that replaced them correspond to later seres, and fully sustainable systems of agriculture form the climax communities of this dimension of human ecology, recycling their nutrients and maintaining stability over millennia.

Factor in the emergence of industrial farming in the early 20th century, though, and the sequence slams into reverse. Industrial farming is an extreme R-selected agriculture; the nutrients it uses come from fertilizers manufactured from natural gas, rock phosphate and other nonrenewable resources, and the crops themselves are shipped off to distant markets, taking the nutrients with them. This one-way process maximizes production over the short term, but strips topsoil, pollutes local ecosystems and poisons water resources. In a world of rapid fossil fuel depletion, such extravagant use of them is also a recipe for spiraling failure. Only the illusion of independence — the belief, anatomized in Chapter One, that human society need not concern itself with ecological limits — makes such shortsighted behavior seem to make sense.

Fortunately, the replacement for this unsustainable system — the next sere in the agricultural succession — is already in place, expanding rapidly into the territory of conventional farming. Modeled closely on the sustainable farming practices of Asia by way of researchers such as Albert Howard and F.H. King, organic farming methods move decisively toward a K-selected model by using organic matter and other renewable resources to replace chemical fertilizers, pesticides and the like. For those people who think of progress in purely technological terms, this looks like a step backward, since it replaces chemicals and machines with composting, green manure and biological pest controls. In terms of succession, on the other hand, it is a step forward toward a truly K-selected agriculture and the beginning of recovery from the great leap backward of industrial agriculture.

succession and technic societies

Map the same way of thinking onto the system of modern industrialism and the parallels to succession appear just as clearly. As it exists today, industrial society can best be described as a scheme for turning resources into pollution as fast as possible. Resource depletion and pollution aren't accidental outcomes of industrialism. They are hardwired into the system: the faster resources turn into pollution, the more the industrial economy prospers, and vice versa. That forms the heart of our predicament. Peak oil is simply one symptom of a wider crisis — the radical unsustainability of a severely R-selected human ecology — and trying to respond to the peak of worldwide petroleum production without dealing with the need to move toward a broader sustainability will simply guarantee that other symptoms will take its place.

For most of a century now, people who have grasped the nature of our predicament have proposed that our civilization needs to start the transition to sustainability deliberately and at once. In the 1970s, in particular, many proposals for the transition saw print, and even today a new one surfaces every year or so. Many are well conceived and would likely work, and even the worst would probably turn out better than the present policy of sleepwalking toward disaster. Still, not one of them – not even in the midst of the 1970s energy crises — received more than brief consideration, either from the government and business interests that make routine decisions in modern societies or from the mass of the population whose opinions form the court of last appeal.

There are many ways to understand the failure of contemporary societies to deal with the onset of the limits to growth. Approach the issue from the standpoint of human ecology, though, and a rarely considered possibility emerges. If the transition between human social systems can be seen as a form of succession, with one society replacing another the way that one seral stage supplants another in nature, then social change may also be shaped and driven by ecological factors, rather than being purely a matter of collective human decisions.

In the succession process in a deciduous woodland biome, for example, grasses replace weeds, shrubs replace grasses and trees replace shrubs in a sequence whose order and time frame can to some extent be predicted in advance. The reasons behind this predictability are deeply relevant to our situation. The bare earth of a vacant lot in Ohio is a welcoming environment for weeds but a forbidding environment for the hardwood trees and other living things that make up the climax community in that bioregion. Pioneer weeds, which thrive on disturbed soil, thus spring up fast and cover the ground. In the process, though, they change the environment and make it suitable, not for more pioneer weeds, but for grasses, and these soon outcompete the weeds and fill the vacant lot in their place.

The same process then repeats itself, as the grasses of the second sere change the environment of the vacant lot and make it better suited to a new sere than to their own descendants. The process continues, gradually slowing down, until it finally reaches the climax community. At this point sustainability has been achieved. The climax community will still change over time, following shifts in climate and the arrival of new species, and it can also be knocked back down to bare earth by a fire or some other disaster, but left to itself it can retain the same recognizable form over centuries. If the quest for a sustainable society parallels the movement of ecosystems in the direction of a climax community, it may not be possible to accomplish it in a single transition.

This succession process parallels the one that took place as human agricultural ecologies spread across the world after the last ice age. Their earliest forms, the R-selected weeds of agricultural ecologies, were well suited to spread fast and establish themselves in regions where the old hunter-gatherer climax communities had been disrupted by drastic climate change. They were not, however, well suited to endure, and were replaced by more K-selected methods of agriculture in turn. It's tempting to suggest that K-selected agriculture could have established itself just as easily in the early days of agriculture if any such thing had been available, but this may well not be true. The farming practices needed to move toward sustainability all require investments of time, labor and resources, as well as knowledge — it takes all of these to build and maintain a compost pile, rotate crops or plow green manure back into the soil — and over the short term, the R-selected methods need less input for the same yield. Only when the R-selected methods fail do arguments in favor of K-selected agriculture begin to look compelling.

Industrialism is following the same trajectory. Like early agriculture, the human ecology of industrialism is an R-selected system, using resources at unsustainable rates. Like early agriculture, its development will be punctuated by crashes; the first of these bids fair to begin in the next few years. It is possible that this crash, or some later one, will spell the end of the entire project — not all new ecologies bear fruit — but it is more likely that more K-selected versions of the same ecology may eventually find their way to sustainability. The logical result of this succession is the rise of a new human ecology that uses high technology produced, powered and maintained with renewable resources.

From the standpoint of the far future, in fact, modern industrialism may turn out to be a primitive and vastly inefficient form of the *technic society*. Like other human ecologies, the technic society can be defined by its energy sources. A hunter-gatherer society uses energy in the form of food, firewood and raw materials taken directly from natural ecosystems. A nomadic herding society also gets its energy from natural ecosystems, but uses livestock as an energy harvesting technology. A village agricultural society does the same thing using domesticated plants. An urban agrarian society uses energy in the form of food from artificial ecosystems created by human labor and supplements this with modest amounts of nonfood energy in the form of fuels, wind, hydropower and sunlight.

A technic society, by contrast, relies primarily on nonfood energy. Modern industrial civilization is simply a form of technic society that gets its nonfood energy from fossil fuels and maximizes production of goods and services in the usual R-selected way at the cost of vast inefficiency. At the other end of the spectrum is the climax community, the *eco*technic society, which gets its nonfood energy from renewable sources and maximizes the efficiency of its energy and resource use in the usual K-selected way at the cost of more restricted access to goods and services.

If this is correct, our own civilization is pursuing a wholly misguided image of what advanced technology looks like. Since the late 19th century, when science fiction writers such as Jules Verne began to popularize dreams of future technologies, "advanced technology" and "extravagant energy use" have been for all practical purposes synonyms, and Star Trek fantasies still dominate discussions of what a mature technological society might resemble. If the model just outlined has any validity, though, a truly mature technology may turn out to be something very different from our current R-selected expectations — and this requires a radical rethinking of most ideas about the future.

As the industrial age ends, the vision of the future that grows from this rethinking may exert a powerful appeal. Still, we are nowhere near the ecotechnic age yet, and if the succession model is any guide, trying to leap directly from the rank weeds of industrial society to the verdant forest of an ecotechnic civilization simply won't work. Even leaving succession aside, we have only the vaguest idea of what a truly sustainable technic society would look like, and history suggests that a long process of trial and error will be needed to develop a technic civilization that can sustain itself for the long term.

In important ways, though, this is simply a restatement of points already made. If human societies replace one another in something akin to ecological succession, the societies that rise from the ruins of industrial civilization will be those best suited to the environment created by their predecessors. These new societies will then be replaced by other societies, until some approximation to a climax community is reached. Only when conditions support the climax community will the ecotechnic future arrive.

the long road to sustainability

The dream of building an ecotechnic society here and now is, of course, widespread. It can be traced in some of the best visionary literature of recent decades, and has been cherished by many people in alternative circles.[5] That dream has become important in some corners of today's cultural dialogue because it embodies a canny strategy for getting past the less productive assumptions that shape contemporary thinking about social change.

Much of the rhetoric used to justify today's social arrangements draws an imagined contrast between the benefits of industrial societies and the supposed squalor and misery of preindustrial life. Many critics of industrialism fall into the trap of accepting the same forced choice while simply reversing the value judgments, as though it's possible to break out of a dualistic way of thinking by standing the dualism on its head. The cleverness of the ecotechnic dream is that it splits the difference by proposing a third option that borrows many of the best qualities from each side. The Hobson's choice between two whole systems, with no alternatives, changes to an open field in which each factor that could make up a future society can be assessed on its own terms. Thus the strategy widens the field of choices, not just to three, but to infinity.

Still, turning this rhetorical strategy into a practical program is harder than it looks. It's popular to think that social change is driven primarily by deliberate human choice, but this is simply another form of the illusion of independence: it assumes that social technology trumps natural limits. The science of human ecology and the evidence of history — and history is simply human ecology mapped onto the dimension of time — both paint a different picture. What they show is that people may attempt to build any society they like, but unless their plans take ecological realities into account, they will fail. Even if a society accepts the hard limits of ecological reality, it will still fail if another society competes for the same resource base more effectively.

The industrial economy now lurching toward history's compost bin, after all, did not achieve global dominance because the people of the world agreed to make that happen. Nor did the squabbling political classes of the world's societies make that decision; there were industrialists who did their level best to further its spread, but there were also powerful people, many members of the old feudal landowning class among them, who staked everything they had on resisting it and failed. Industrial civilization had its day in the sun because, in a world where fossil fuel could be had for the digging or drilling, the industrial mode of production was more efficient than its rivals, and enabled the communities that embraced it to prosper at the expense of those that did not. As the industrial system undercuts the environmental conditions that allow it to thrive, new forms better adapted to the new reality will elbow today's industrialism aside and take its place. We have our preferences, but nature has the final say.

Apply the same measure to the rise of an ecotechnic society and the challenge is clear. The conditions that would make an ecotechnic society the most successful option are roughly those that existed before the industrial revolution broke open the Earth's fossil fuel reserves and started looting them for short-term advantage. In a world where the available energy resources are sun, wind, water, muscle and biomass, and all work must be accomplished by those means, societies that evolve efficient and sustainable technologies using those resources have major advantages against rival societies that use them unsustainably.

The problem with building an ecotechnic society here and now is that the conditions just outlined do not yet exist. So far, humanity has used around half the world's stock of petroleum, and a little less than half its stock of coal and natural gas. These fuels will be available in diminishing amounts for a long time to come. While modern industrial societies as they exist today probably can't survive the end of cheap energy, peak oil is already driving the emergence of *scarcity industrialism*, a new human ecology better adapted to a world of dwindling fuel supplies. While fossil fuels can still be produced in useful amounts, scarcity industrialism will likely produce more wealth and exert more power than any ecotechnic system. Societies with fossil fuels have historically overwhelmed those without them, and nothing suggests this will change soon.

In the longer run, a second new ecology, the salvage society, is likely to replace scarcity industrialism in its turn. Many relics of today's industrial societies will still exist far into the future. These legacies represent stored energy — they embody the energy that was needed to create them, and to build the material and knowledge base that made them possible — and the additional energy needed to maintain and use them in many cases will be quite small compared to the stored energy contained in them. The energy needed to keep a hydroelectric plant or a computer in working order is fairly small compared to the energy they embody, or the advantages that owning and using them could confer.

It's quite likely, therefore, that deindustrial societies that can no longer build a hydroelectric plant or a computer could still maintain the less demanding knowledge and resource base needed to keep them running, in the same way that Dark Age societies all over Europe used and repaired Roman aqueducts they could never have built themselves. The resulting salvage societies will have advantages that purely ecotechnic societies will not, and so these human ecologies will spread wherever the supply of potential salvage allows them to function. Still, their time will pass; many of the legacies of the industrial age will not be renewable, and when they're gone, they're gone.

The result is a striking parallel to succession. In the near and middle future, as the deindustrial age unfolds, the societies that will flourish best are those that will be least able to survive over the long term. In the near term, societies that embrace scarcity industrialism, relying on efficient use of remaining fossil fuels and eking them out with renewable resources and high technology, will likely do better than either the wasteful abundance economies of the present or the more sustainable cultures that will replace them. In the middle term, salvage societies that combine sustainable subsistence strategies and economies with effective use of the industrial age's legacy technologies will likely do better than the lingering fossil fuel-using societies they replace, or the ecotechnic societies that will replace them in turn. Only when coal and oil are rare curiosities, and the remaining legacies of the industrial age no longer play a significant economic role, will ecotechnic societies come into their own.

It's crucial to keep this process in mind when planning for the end of the industrial age. The last years of today's economy of abundance, the decades of scarcity industrialism built on the last significant supplies of fossil fuels, and the century or two of salvage societies in the middle future, form three hurdles that have to be leapt in order to get to the ecotechnic age. Instead of trying to make the leap to an ecologically balanced, fully sustainable society all at once, the transition must be made one hurdle at a time, adapting to changes as they happen, and trying to anticipate each seral stage in time to prepare for it, while working out the subsistence strategies and social networks of the future on smaller scales.

This approach is evolutionary rather than revolutionary — that is, it relies on incremental changes and continuous experimentation, rather than trying to impose a rigid break with the past and an ideological pattern that may turn out to be less viable that the one it replaces. This is necessary because the human ecology that succeeds best under any set of environmental conditions depends much more on those conditions, and the way they interact with available resources and technology, than on choices we make. Constructive changes are possible in almost any situation, but only within the limits imposed by ecological realities, and societies that try to ignore those realities will face stark handicaps in facing the challenges of survival and competition from other, less burdened societies.

Nobody alive today knows what a truly sustainable technic society would look like, much less how to build one. The only form of technic society human beings have yet experienced is the industrialism of the last 300 years, and nearly everything that made that system work will be gone once the age of cheap abundant energy ends. The time of contraction ahead of us is, among other things, an opportunity for social evolution, in which various populations will try out many different forms of technical, economic and social organization, some of which will turn out to be more successful than others.

Out of that process will evolve the successful ecotechnic societies of the far future, perhaps three centuries from now, perhaps more. The journey there, however, will be made more challenging by the impact of today's choices on the future taking shape around us.

A Short History of the Future

IN THE FUTURE, as John Maynard Keynes famously pointed out, all of us will be dead. That makes the present more interesting to most of us! Given the crises we will likely encounter in the next decades, it's understandable that many people focus on current problems and let the distant future take care of itself. Still, the importance of the outcome makes a glance at the long view worth taking. The further an archer plans on shooting, to extend a metaphor from Machiavelli, the higher he needs to aim, and the further downrange he needs to track his target.

Still, Machiavelli's archer had the advantage of being able to see his target. Those of us who try to glimpse the future have no such luck. Even when the broad sweep of events can be predicted, the details usually head in unexpected directions. Many people predicted the First World War, but nobody dreamed that it would turn a penniless exile who wrote under the pen name "Lenin" into the Communist dictator of Russia and topple Nicholas II from what most people thought was the most secure throne in Europe. Surprises on the same scale are doubtless lying in wait in our own future.

In his valuable book *The Black Swan*, Nassim Nicholas Taleb showed that most of the dominant facts of contemporary life have been shaped by such surprises. He pointed out, for example, that no one could have known that Google, which began as one Internet search engine among many, would rise to dominate much of the Internet, while other equally promising firms went under.[1] He's quite correct, but there's another side to the story. Anyone familiar with the history of technology could have predicted that one company in the field would break out of the pack and dominate the market, because this consistently happens with new technologies. Google's path to dominance followed the same course as Microsoft's, say, or RCA's emergence as the powerhouse of the American radio industry in the 1920s. No one could have known in advance *which* search engine would dominate the market, but it was very nearly a foregone conclusion that one would.

In the same way, the details of the economic crisis of 2008 were impossible to guess in advance, but many people noticed that the freewheeling speculation that caused it matched the excesses that led to previous economic crises. Long before the US housing market began to crash, that awareness had crystallized into public predictions that massive financial crisis was in the offing unless banking and investment practices changed drastically.[2] Those predictions were utterly correct; no one could be sure when those crises would hit, or what their immediate causes would be, but the history of financial crises made predicting another one fairly easy.[3] Thus prophecy works best when it pays close attention to history and uses the past to gauge the kind of events that will occur in the future, rather than trying to predict the events themselves.

It's popular these days to insist that because the changes hitting industrial society are more drastic than anything in recent memory, they are more drastic than anything in history, and the past can therefore be disregarded as a guide to the future. This argument seems hopelessly misguided to me. Just as Google's rise tracked the emergence of other successful companies in fields dominated by new technology, and the housing bubble followed the same course as many other examples of speculative booms and busts, many civilizations before ours have gone into overshoot and traced out the trajectory of decline and fall. Our civilization has already followed the early stages of that trajectory. People today like to think that their situation is unique and their society is therefore exempt from the common fate of civilizations, but the people of many other civilizations have made the same claims about their own uniqueness, too.

glimpsing the deindustrial age

More dramatically than any other form of social change, the fall of a civilization has sweeping impacts on human life, and an entire book could be devoted to the subject. Here I want to discuss four of these impacts that will likely shape the future and have particularly strong effects on the transition to the ecotechnic age. All of them unfold from changes already at work today.

1. **Depopulation.** People nowadays are so used to worrying about the population explosion that the possibility of its opposite rarely comes up for discussion. Yet the population bubble of the last few centuries is just as much a product of the exploitation of fossil fuels as the industrial age itself. Without changes in agriculture, trade and public health launched by the needs of a fossil fuel-powered society, the relatively modest surge in human numbers in the 19th century would have reversed itself in the normal way. In point of fact, it nearly did so; at the dawn of the 20th century, bubonic plague surged out of central Asia, and only massive efforts by the colonial powers of the age prevented a third plague pandemic from sweeping the globe.[4] As the industrial age ends, the barriers that prevented pandemics will likely fail, while the food surpluses that support today's population levels will be impossible to maintain. When civilizations decline and fall, their population decreases to a fraction — and sometimes a very small fraction — of their peak level. Our experience is likely to be similar.

2. **Migration.** Depopulation moves at different paces in different cultures and regions, and these and other factors can cause people to leave their homes and head elsewhere. Even in the absence of modern transportation, they can end up far from the places they started. Before Rome fell, for example, the ancestors of today's English lived

in Denmark, the ancestors of today's Hungarians lived in central Asia, and the ancestors of today's Spaniards lived north of the Black Sea. Today, as tidal flows of refugees press at borders worldwide, the only thing preventing equal migrations is the fraying fabric of national sovereignty, backed by military forces dependent on fossil fuels. As the industrial age ends, those bulwarks will fail, and epic migrations will likely result.

3. Political and cultural disintegration. Such migrations assume the collapse of current political arrangements over much of the world, but this is a safe bet. When civilizations come apart, their political systems and cultural traditions either vanish or survive as a shell of formalities covering drastically changed realities. The shell can exert a potent influence of its own — in Europe after the fall of Rome, as in China after the collapse of the Han dynasty, the title of "Emperor" retained immense power even when nobody existed who could claim it, and the gravitational attraction of the old imperial state helped drive efforts toward unification many centuries later. Along the same lines, warlords of the future may well claim the title of President of the United States, centuries after the office and the nation it once served exist nowhere outside of legend.

Culture faces similar changes. Right now the manufacture of mass culture imposes a thin shell of cultural similarity across English-speaking North America, but even that is under strain as subcultures use the decentralizing power of today's communications and move more boldly in their own directions. The end of the industrial age will bring down the Internet, but it will also play taps for the technologies that make mass culture possible. In the bubbling cauldron of a future North America, today's cultural initiatives will fuse with older traditions, imports from abroad and new movements in ways we can't imagine today. The loss of political unity and the end of easy long-distance travel will make the rise of new local and regional cultures certain.

4. Ecological change. Natural systems form the foundation of all human societies, and the fall of civilizations in the past has often followed major ecological shifts. The sheer scale of industrial civilization, though, gives the consequences of its heyday and decline far wider impact than past civilizations, and will likely set ecosystems spinning into new patterns over much of the globe. Climate change is only one aspect of this picture, though its importance should not be understated. Ecosystems are complex enough and change over such varied timescales that many of the effects of industrial civilization's rise and fall are challenging to predict, and a great many more will be known only after they happen.

Many other factors will play important roles as well, including some that can't possibly be anticipated here and now. Still, these four trends promise to shape the future in crucial ways, and impose significant constraints on the long journey to the ecotechnic future. The pages that follow will examine each trend in more detail.

the depopulation explosion

A basic fact of our predicament is the hard reality that today's human population is far larger than the world's carrying capacity. What William Catton called "ghost acreage"—the vast boost to subsistence that fossil fuels give to growing, storing and distributing food[5]—has allowed the world's human population in the last few centuries to balloon to disastrously high levels. As the industrial age ends, the surpluses of food and other resources and the infrastructure of public health that supported this expansion will end as well, with predictable impacts on the size of the human population.

The resources supporting industrial civilization, however, can be expected to taper off rather than coming to an abrupt end. Petroleum, the most important of today's energy resources, is a case in point. The Hubbert curve—the bell-shaped curve devised in the 1950s by petroleum geologist M. King Hubbert[6]—shows that roughly half the oil in any resource, from a single well to a planet, will be pumped out of the ground after the peak of production arrives and passes. Production rates along the downside of the curve are also roughly commensurate with production rates at points on the upside the same distance from the peak. If the peak of Hubbert's curve remains at 2005, in other words, the amount of oil produced in 2030 will likely be somewhere near the amount produced in 1980; production in 2060 will fall near production in 1950, and production in 2100 will approximate production in 1910. Even after the peak comes and goes, in other words, there will still be a great deal of oil for many years to come. The same rule applies to other energy resources.

This lesson could have been learned from the growth of nonconventional oil sources like the Alberta tar sands, and the reopening of hundreds of formerly abandoned stripper wells in pumped-out oil regions like Pennsylvania. As oil production falters, market forces and political pressures guarantee that every possible replacement will be brought online. Right now, attempts to increase production are struggling to keep up with slumping yields at existing fields, and it's a struggle that will only get harder as more fields reach their own Hubbert peaks. Still, even though new fields and alternative sources can't make up for the exhaustion of existing fields, they can stretch out the decline substantially. The fact that the "ghost acreage" that supports today's huge global population is going away gradually, rather than all at once, does not change the fact that it's going away. This alone makes a decline in global population inevitable.

It doesn't take much to change an expanding population into a contracting one. Modest changes in birth and death rates will do the trick, and such changes are predictable consequences of the twilight of the industrial age. The factors that push population contraction in hard times are familiar enough to demographers: malnutrition, epidemic disease and child mortality driven by failing public health, and social factors such as alcoholism, drug abuse, violence and suicide, driven by the psychological impacts of life in a disintegrating society.

A preview of the future is already showing in Russia and other fragments of the former Soviet Union where crude death rates have risen to nearly double the rates of live births, a trajectory that will cut population figures in half by 2100 or so. Similar population contractions can be traced in the declining phase of many past civilizations. As the industrial age unwinds, similar patterns will likely unfold in North America; for that matter, whole regions of the American West are depopulating right now. Archeologists of the future may well mark the beginning of ancient America's decline and fall with the failure of settlement on the western plains in whatever the late 20th century works out to in some future calendar.

As this suggests, the process of depopulation will be powerfully shaped by geographic factors. Communities that are economically viable in a global economy awash in cheap fossil fuels, in many cases will not be economically viable in the deindustrial future. This cuts both ways. Sprawling Sun Belt cities with little water and no resources will shrivel and die as the energy that keeps them going sputters and goes out, and tourist communities across the continent will pop like bubbles and become ghost towns once travel becomes a luxury, while Rust Belt towns struggling for survival today will likely find a new lease on life when adequate rain, workable soil and access to waterborne transport become the keys to prosperity, as they were in the early 19th century.

völkerwanderung

What is less certain is whether it will be the descendants of today's Americans or some other peoples who will populate the renewed Rust Belt towns and salvage valuable metal scrap from the crumbling ruins of today's Sun Belt cities. Mass migration is already a fact of life throughout the contemporary world, and the twilight of cheap energy promises to shift this into overdrive.

It's common today to think of nations as a fixed reality with which historical changes have to deal, but this is far from true. Even in periods of relative stability, populations move, cultures relocate and nations flow, fuse and break apart like grease on a hot skillet. A hundred years ago the Austro-Hungarian Empire and the Ottoman Empire were major players in world politics — try finding either one on a map today. Norway had just won its independence from Sweden, the thought of West Indian or Pakistani communities in Britain would have drawn blank stares, while debates over immigration in the United States focused on whether Italians were white enough to be welcomed. People at that time, like most people today, imagined that the state of affairs familiar to them would continue into the indefinite future. History proved them wrong, and it is unlikely to be any kinder to our own assumptions of permanence.

German historians of the 19th century coined a useful word for the age of migrations that followed the fall of Rome: *Völkerwanderung*, "the wandering of peoples." Drawn by the vacuum left by the implosion of Roman power and pushed by peoples from the steppes further east who were driven westward by climate change, whole nations took to the road. It took a thousand years before the migrations settled into a stable pattern, and by the time that happened very few of the peoples of Europe were living in the same places as before. This has happened many other times in the past when empires came apart. What makes it important here is that we are likely to see a repeat of the phenomenon on a grand scale in the near future.

The first ripples of the future flood can be seen by anyone who travels by bus through the rural United States anywhere west of the Mississippi River or south of the Mason-Dixon line. Stray from the freeways and tourist towns and culturally speaking, as often as not, you're in Mexico instead of the United States: the billboards and window signs are in Spanish, advertising Mexican products, music and sports teams and people on the streets speak Spanish and wear Mexican fashions. It's popular among Anglophone Americans to think of this as purely a Southwestern phenomenon, but it has become just as common in the Northwest, the mountain states and large sections of the deep South. There are some 30 million people of Mexican descent in the US legally and some very large number —

no one agrees on what it is, but eight million is the lowest figure anyone mentions — who are here illegally. As the migration continues, much of what was once the United States is becoming something else.

A great deal of angry rhetoric has flared from all sides of the current debates on immigration, but none of it deals with the driving force behind these changes — the failure of the American settlement of the West. The strategies that changed the eastern third of the country from frontier to the heartland of the United States failed to work west of the Mississippi. Today the cities and farm towns that once spread across the Great Plains are fading into memory as their economic basis vanishes and the last residents move away, while the mountain and basin regions further west survive on tourist dollars, retirement income or cash crops for distant markets — none of them viable once cheap energy becomes a thing of the past.

Like the Mongol conquest of Russia or the Arab conquest of Spain, the American conquest of the West is proving to be temporary, and as the wave of American settlement recedes, the vacuum is being filled by the nearest society with the population and cultural vitality to take its place. The same thing is happening in Siberia, where Chinese immigrants stream across a long and inadequately guarded border, making the Russian settlement of northern Asia look more and more like a passing historical phase. Such shifts are very common when the reach of a powerful nation turns out to exceed its grasp.

Once again, such changes shift into overdrive when civilizations break down. In an age of disintegration, when the political and military power that backs up America's borders will likely come unraveled, and climate shifts could hand tens of millions of people in Latin America a Hobson's choice between migration or starvation, völkerwanderung becomes probable. Map the Roman model onto the present and it's conceivable that by the year 2500 or so, the people living in today's Iowa and Wisconsin might trace their origins to a migration from Brazil, while west of the Mississippi, languages descended from English might only be spoken in a few enclaves in the Pacific Northwest.

History also shows that where technology permits, völkerwanderung can easily leap oceans. From the sea peoples who ravaged the eastern Mediterranean world around 1300 BCE to the Vikings of the early Middle Ages who sailed to Greenland, Russia and Sicily, many migrant peoples have taken to the sea. As the petroleum age winds down, many nations with large populations, limited resources and strong maritime traditions will have few options other than mass migration by sea. Consider the situation of Japan: close to 150 million Japanese now live on a crowded skein of islands with little arable land and no fossil fuels at all, supported by trade links made possible only by abundant energy resources elsewhere. As fossil fuel production declines, industrial agriculture and food imports both will become problematic, and over the long term the Japanese population will have to contract to something like the small fraction of today's figures the Japanese islands supported in the past. Mass migration is nearly the only option for the rest of the population. Japan's ample supply of ships and fishing boats provide the means, and possible destinations beckon all around the Pacific basin.

The possibility that twenty million Japanese "boat people" could follow the Pacific currents to the west coast of North America by 2075 or so, or that millions of Indonesians might head for the northern shores of Australia for the same reason, has not yet become a part of our collective discourse about the future. Our unwillingness to grapple with the likelihood of mass migrations in the wake of the industrial age, however, will do nothing to make the impacts of population shifts easier to face.

culture death

The political and social landscape of the industrial world may not need mass migration to face dramatic change, however. The industrial age has also been the age of the nation-state. In a cascade of change that began a century before the industrial revolution, nation-states defined themselves on two fronts: against both local loyalties and the transnational community of Christendom. People who thought of themselves as Cornish or Poitevin or Westphalian, on the one hand, and members of the universal Body of Christ on the other, struggled to cope with new social realities that demanded they think of themselves as English or French or German.

The nation-state as a source of identity depended on a deliberate blurring of categories in which a population, a culture, a language and a system of government fused in the collective imagination into a single entity. One of the consequences of this category blending is that populations, cultures and languages become gaming pieces in the struggles of local power centers to define and defend themselves against national governments. Watch the long struggle over the Welsh language in Britain, for example, and you have a ringside seat for the conflict between centralizing and decentralizing forces in British political life. The notion of race had similar origins as members of multiethnic societies tried to define themselves in ways that excluded economic or political rivals.

These issues have special relevance today, because the nation-state has drawn most of its strength from the economic and political integration of Western nations over the last three centuries, and this has been inseparable from the rise of an industrial economy powered by fossil fuels. It's not accidental that Britain, the first nation-state to break through to industrialism, was also one of the first Western states to form a coherent national identity. The transportation networks that made industrialism work — canals, then railroads, then highways — also made it possible for national governments to extend their reach throughout their territories in ways few previous societies ever managed.

The history of North America provides a good example of the process. In 1861 it was still possible for many people in the South to think of themselves primarily as Virginians or Georgians or Texans, and only secondarily as citizens of the United States. Sixty years later, even the Ku Klux Klan had to define its repellent goals as "100 percent Americanism" in order to find an audience. In 1861, the North American railroad network was still in its infancy, mostly concentrated in the Northeast and Midwest. By 1921 it blanketed the continent with one of the most successful transportation systems in history and was already being supplemented with highways and airplanes. As transport expanded, so did the reach of the federal government, and so did the focus of most Americans' sense of identity.

It's been common for believers in progress to argue that national governments will go the way of the feudal provinces and half-independent states that were swallowed up by the growth of the nation-state, and give way in turn to a global government. They might well be right, too, if we could count on an ever-increasing supply of cheap abundant energy, but of course we can't. Rather, as energy becomes scarce and expensive, transportation networks that depend on cheap abundant fuel will unravel. Those countries with good railroad systems will likely be able to maintain them after the highways are silent, and the networks of last resort, the canal systems that made 18th century industrialism work, remain viable in many European countries and may put a floor under decline if their value is recognized in time. Most European countries are also built on a geographical scale that makes sense in a world of limited travel. Unless migration overwhelms them, their chances of remaining intact in the deindustrial future are tolerably high.

The United States, by contrast, scrapped most of its world-class rail system in the third quarter of the 20th century, and few sections of its old canal system still survive today. Easy travel and mass media, at this point, provide most of the glue that holds the United States together, and the breakdown of that glue will likely see the unraveling of the United States as a focus of collective identity. Just as the rapid growth of transportation links turned the grandchildren of Virginians and Californians into Americans, the disintegration of those links may turn the grandchildren of Americans into something else.

It's unlikely to turn them back into Virginians and Californians, though, because the triumph of the nation-state in the 19th century was followed in the United States, more than anywhere else, by the triumph of the market economy over culture. A faux mass culture designed by marketing experts, produced in factories and sold over the newly invented mass media, elbowed aside the new and still fragile national culture of the United States and then set to work on the regional and local cultures the latter had just begun to supplant. By the second half of the 20th century, the tunes people whistled, the recipes they cooked, the activities that filled their leisure hours and the images that shaped their thoughts and behavior no longer came out of such normal channels of cultural transmission as family and community; they came out of the market economy, with a price tag attached that cannot be measured in dollars alone.

This period, in fact, saw the death of anything that could reasonably be called American culture. Most examples of what anthropologists call "culture death" have seen people beaten and starved into giving up their traditional cultures; what the American experience shows is that people can also be bribed by prosperity and cajoled by advertising into doing the same thing. In a society where media campaigns take the place of less mercenary guidance, making traditional American culture look bad was just another bit of marketing. Think of the movie "Deliverance," with its likeably cosmopolitan heroes struggling to survive against the brutal malevolence of backwoods hicks; the banjo riff that provided the movie's leitmotif defined traditional American culture itself as a hostile Other.

Culture death is a traumatic experience, and I suspect that a great deal of the shrill anger and self-pity that fills American society these days has its roots in our unwillingness to face up to this trauma. As the age of cheap energy comes to an end, though, I suspect there are worse traumas in store. A nation that has sold its culture for a shiny plastic counterfeit risks a double loss if that counterfeit pops like a soap bubble in its collective hands. Equally, a people that has come to see itself as a passive consumer of culture, rather than its active maker and transmitter, may have very few options left when the supply of manufactured culture runs out.

The impact of these dilemmas on our collective imagination of identity is likely to be drastic. We are already seeing people in contemporary American society turn to any imaginable resource in the search for some group identity less transient than the whims of marketers. Religion has filled its time-honored role in this regard, but so have racial fantasies, sexual habits, social theories and more — speakers of Klingon or J.R.R. Tolkien's Elvish, for example, outnumber speakers of quite a few real languages. This is still a fringe phenomenon, though less so than it was twenty years ago; twenty years from now, such subcultures may impact the mainstream in ways impossible to predict today.

What we can predict with a fair amount of assurance is that the cultural patterns that will rise in industrialism's wake will have little in common with those that exist now. As the mechanisms that keep mass culture going break down, cultural legacies from before the age of mass media, contemporary subcultures, immigrant cultures and emerging cultural initiatives will find themselves in the same ferociously Darwinian environment that followed the collapse of Rome or the decline and fall of any other cosmopolitan civilization. Those cultural traditions and practices that foster survival will endure; those that do not will vanish. Today's cultural smorgasbord, in effect, will be plundered to fill a wide range of bubbling stewpots, and the long slow cooking that follows will leave few of the ingredients unchanged.

a different planet

All these transformations and more will be powerfully influenced by the impact of today's industrial human ecology on the biosphere. One dimension of that impact — the unrestricted dumping of greenhouse gases from tailpipes and smokestacks over the last few centuries — has received recent attention, and for good reason. Climate, more than any other factor, determines the ecological possibilities anywhere on Earth's surface; given the annual rainfall and minimum and maximum temperatures for a given location, a good ecologist can usually predict what form of climax community will appear there. Change the climate of a region and every dimension of that region's ecology will shift; change the climate of the Earth and the results redefine the biosphere. This is what industrial civilization is doing by dumping billions of tons of carbon dioxide and other heat-trapping gases into the atmosphere.

This is serious enough. At the same time, estimates of global climate change have not always started from a realistic sense of industrial society's limits. For example, the International Panel on Climate Change (IPCC) — the body of scientists charged with assessing the risks of climate change — takes it as a given that world production of petroleum, coal and natural gas will increase steadily through the year 2100.[7] This is problematic, to say the least: conventional petroleum production has already peaked, natural gas is expected to peak around 2030, and by 2040, according to several cogent studies, coal production will have peaked as well.[8] By 2100, then, consumption of all fossil fuels put together will be a very modest fraction of today's levels, because there will be very little in the way of fossil fuels available. The resulting climate changes will thus fall far short of the doomsday scenarios circulated on the far end of the global warming activist community.[9]

Still, this hardly makes global climate change irrelevant. The results of climate shifts in the past included crop failures, major shifts in rain and temperature belts, and significant changes in sea level. Even very modest shifts have been enough to tip civilizations over the edge into permanent decline.[10] A great many of these changes are already in the pipeline due to greenhouse gases already dumped into the atmosphere, and since no national government on Earth has shown a willingness to accept the huge economic and social costs of a retreat from fossil fuel consumption, there seems little likelihood that such events can be prevented.

The same dynamic governs other ecological transformations already under way. Perhaps the most significant of these is the massive reshuffling of species going on across most of the world. Some species are at risk of extinction, or have already gone extinct, while many more have had their geographical distribution completely transformed—decreased in some cases, increased in others—by deliberate or accidental human intervention. The spread of exotic plants and animals into new bioregions is shuffling the planetary ecology like a pack of cards. Add climate change and other forms of environmental stress, and evolutionary pressures shift into overdrive, guaranteeing that many of the ecosystems our descendants will encounter will have little in common with the ones around us today.

In effect, human beings in the deindustrial age will live on a different planet. It will be warmer than our world and much of it will be wetter as rising temperatures cycle water vapor up from the oceans and increases the planet's total rainfall. Climate belts will shift poleward, changing the crops that can be grown in most of the world's agricultural regions. Coastal regions within fifty feet or so of sea level will drown beneath rising seas. The natural ecosystems of this future Earth will undergo something like a planet-wide succession process, setting the stage for significantly new ecosystems over much of the planet. All of this forms the setting in which the societies of the future will evolve.

the world is round

If history is any guide, none of these patterns will affect the whole Earth in the same way. Population decrease, like population increase, is influenced by geography and culture; migrations sweep across some areas but leave others unaffected; political and cultural shifts are exquisitely sensitive to local factors; and the global biosphere is made up of hundreds of bioregions and countless individual ecosystems, each of which will be affected by ecological shifts in its own way. In the short run, this will make many trends easy to miss in those areas little affected by them and easy to exaggerate in the regions that bear their brunt. In the long run, it guarantees that the future will take many forms.

In his bestselling 2005 book, *The World Is Flat*, Thomas Friedman used the metaphor of a flat Earth to suggest the compression of distance that has been so potent a part of history in the 20th and early 21st centuries. Linked by telecommunications and cheap air travel, Earth's nations and cultures have all been within easy reach of one another for the first time in recorded history. The result has been a flattening out of regional, national and continental differences. The end of the age of cheap energy promises to throw that process into reverse. As the world becomes round again and cultural and economic exchanges across continents slow to a trickle, human populations will go their own ways and make their own histories.

While it's unlikely that the continents will ever be as isolated from one another as they were before 1492, there may be long centuries ahead when the only news about the doings of other lands will take the form of travelers' stories and tall tales in seaport bars. Nor will the societies of the future necessarily share a common technological basis or human ecology, as many of them do today. A few centuries from now, it is entirely possible that some regions may support urban centers and a relatively advanced technology while other regions are inhabited by village agriculturalists, nomadic herding peoples, hunter-gatherer tribes or cultural patterns that do not yet exist.

It is important to keep in mind that these changes will not happen all at once. Just as today's technologies foreshorten distance, modern ways of thinking about history tend to foreshorten the time frame of historical change. It becomes easy to forget that no matter how drastic a change appears on the pages of history books, it is rarely sudden for those who live through it. Read a history of the French Revolution and events seem to follow one another like a string of exploding firecrackers from the final crisis of the ancien régime until the fall of Napoleon. For the man or woman in the French street, though, these happenings were scattered threads in a fabric of months and years woven from the plain cordage of ordinary life.

A teenage Parisienne who sat daydreaming of her upcoming wedding on the day that Louis XVI summoned the États-General in 1788 would have been a grandmother on the day the Allied armies marched into Paris after the battle of Waterloo in 1815. Historical events rarely appear to have the same importance at the time that they are assigned in the historian's hindsight, not least because the everyday tasks of making a living and the stages of human life play a larger role for most people than the tumults that make the history books.

Seen in retrospect, the changes that will follow the end of the age of cheap abundant energy are likely to be swift and global. From the perspective of those who live through them, though, those changes will take place over lifetimes, and will be powerfully affected by local and temporary factors that will make the broader trends harder to see. As petroleum production declines, for example, the scramble for replacements with less net energy will tend to drive fuel prices up. Economic contraction driven by the end of cheap energy will tend to decrease demand, and drive them back down. Factor in speculation, and in place of the steadily rising prices sometimes predicted in the peak oil literature, we can expect wild swings in energy prices, driving cycles of boom and bust of an intensity not seen in the Western world since the 19th century.

All of this spells trouble, without a doubt. To volatile energy prices and wrenching economic change, add the breakdown of public health and the likelihood that the end of the American empire will result in wars as bloody as those that followed the decline of every other empire in history, and the result is a recipe for massive change. As depopulation, migration, cultural drift and ecological change have their effects, those changes will be multiplied manyfold. From the perspective of some future Edward Gibbon of the year 3650 or so, outlining *The Decline and Fall of the American Empire* as he strolls past sheep grazing on the mossy ruins of ancient Washington DC, all this will doubtless seem traumatic enough.

For those who experience the process of transformation first hand, though, it will likely have a noticeably different appearance. The young Parisienne just mentioned, after all, did not go to sleep one night in the agrarian, half-feudal France of the ancien régime and wake up the next morning as a grandmother in the nascent industrial nation that France became in Napoleon's wake. Even those changes that brought grief into her life — any sons she had, for example, would have faced high odds of dying a soldier's death — would have been spread out over the years, part of a fabric of many other experiences.

Similarly, the unraveling of today's industrial society can be expected to take place against the tempo of ordinary life. Those of us who live through any significant fraction of the process can expect to witness economic, social and political turmoil as dramatic as anything our ancestors have experienced. We will all be attending more funerals than most of us do nowadays, and our appearance as the guest of honor at one of them will likely come sooner than we expect. Most of us will learn what it means to go hungry, to work at many jobs, to watch paper wealth become worthless and to see established institutions go to pieces around us. A quarter century or so from now, the world may be a very different place, but the journey there will have taken place a single day's changes at a time, in a world far more diverse than the one we inhabit today.

Toward the Ecotechnic Age

NONE OF THE possibilities just outlined are unusual in human history. Until recent times, population loss, mass migration, major political and cultural shifts, and devastating ecological changes were routine events. Only the fact that most people in the industrial world have not personally experienced such events makes them seem unlikely today. The dubious foundations underlying our conviction of our own uniqueness show themselves here with remarkable clarity.

The same sense of uniqueness drives the belief, a widespread one nowadays, that the rising spiral of troubles facing industrial civilization herald an evolutionary leap that will give rise to a new world lacking most of the problems that beset the present one. This belief system began to take shape in the late 19th century, shortly after evolution itself entered our cultural dialogue, and every crisis from that time to this has given rise to proclamations that the evolutionary leap was about to happen. The cultural changes of the 1960s, however, gave such claims a wider presence in contemporary society, and in recent years books and speakers announcing the arrival of a new stage of evolution have been guaranteed a substantial audience.[1]

Central to all these claims are a set of assumptions about the way evolution works. These assumptions hold that evolution moves in a specific direction by way of clearly defined stages that can be set in order from lower to higher; that organisms at these higher stages are more evolved than those at lower stages; that human societies can likewise be ranked in linear order from less evolved to more evolved forms; and that we can therefore expect industrial society to be replaced by a more evolved form, which will have more options and fewer problems than today's. All these beliefs are widely held in today's culture; their only drawback as a guide to the future is that every one of them happens to be wrong.

The mismatch between today's popular notions about evolution and the realities of evolution as understood by biologists, is among the debris sent flying by the head-on collision between Darwin's theory of evolution and the social obsessions of the era in which that theory emerged.[2] The English caste system had claimed religious backing for centuries; as religious ideas lost support in the 19th century, scientific justifications for social hierarchy became a growth industry. By the time the ink was dry on the first copies of *The Origin of Species*, evolution had been pressed into service in this dubious cause. The centerpiece of these efforts was a claim that foreshadows George Orwell's satire *Animal Farm*: just as Orwell's ruling elite of pigs decreed that all animals are equal but some are more equal than others, popular misstatements of evolution argued that all living things evolve, but some are more evolved than others.

Biologically speaking, this is nonsense. A human being, a gecko, a dandelion and a single-celled blue-green alga are all equally evolved — that is, they have all been shaped to an identical degree by the pressures of their environments. We think of human beings as more evolved than blue-green algae because 19th-century Social Darwinists such as Herbert Spencer practiced conceptual sleight of hand, redefining evolution's outward surge of life toward all available niches as a ladder of social status. The concept of evolutionary stages or levels was essential to this conjurer's act, since it allowed social barriers between English classes to be projected onto the inkblot patterns of the biological world. The result of this manhandling of biological realities, though, is a complete misunderstanding of the evolutionary process.

what evolution means

Like population cycles and succession, evolution is a natural process that follows ecological laws. It has no stages or levels; it simply has adaptations. There is no straight line of progress along which living things can be ranked and neither individual species nor the biosphere as a whole have changed in any single direction over the course of evolutionary time. Rather, the biosphere has waxed and waned in biological richness and complexity as the Earth's climate has moved through ages more or less favorable to life, while lineages of living things have splayed in all directions like the branches of an unruly shrub. Sometimes those branches took unexpected turns and opened up new possibilities, but these evolutionary breakthroughs can no more be ranked in an ascending hierarchy than organisms can.

This can best be understood through an example, and one that is particularly relevant to humanity's present situation is the evolution of the first bats.[3] Sixty million years ago, in the Paleocene epoch, the flightless ancestors of bats were insect-eating, tree-dwelling nocturnal mammals, related to the shrew-like early primates who launched the evolutionary lineages that led to us. To animals that live in trees, falling is a constant risk, and thus a constant source of evolutionary pressure. Adaptations that will help them cope with that danger have a good chance of spreading through a population; this is how sloths ended up with long claws, New World monkeys got prehensile tails and many animals developed folds of extra skin that function as a parachute if the animal falls out of its tree.

If the extra skin bridges the gap between forelegs and hind legs, as it usually does, the species evolves the ability to glide, like flying squirrels, colugoes and the like. This is viable but it offers few options for further development. A species that evolves extra skin on and around the forelimbs, though, has just made an evolutionary leap into a new world. The muscles of the forelimbs allow gliding in a much more controlled way and it becomes possible to put muscle into the movements — in other words, to fly. Once an insect-eating animal can do something more than a controlled fall, furthermore, every winged insect is potentially on the dinner menu and slight improvements in flying skills pay off handsomely in access to previously unavailable food. This is what happened to the first bats, just as it happened to the first pterodactyls some 200 million years earlier.

Unlike the evolutionary leaps imagined by so many alternative thinkers today, however, the leap into the sky made by the first proto-bats solved far fewer problems than it created. Flight committed them to risks far beyond anything their ancestors encountered scampering around in trees, even as it gave them access to a huge and previously inaccessible food supply. The combination of great risks and great opportunities subjected them to ferocious evolutionary pressure in the direction of efficient flight. Under that pressure, a few hundred thousand generations — an eyeblink of evolutionary time — was enough to turn the first awkward flutterings into agile flight.

By 55 million years ago, as a result, bats almost identical to today's insect-eating varieties were flitting through the Paleocene skies. The sonar that now guides bats to their prey took a while longer to evolve, and some offshoots of the family — the fruit bats and flying foxes, for example — took longer still, but the basic adaptations were set, and to the discomfiture of mosquitoes and moths they have remained viable ever since. As evolutionary breakthroughs go, this leap into flight was a massive success; bats are the second most numerous of the mammalian orders, exceeded only by the even more successful rodents. Still, it takes an even more breathtaking leap of logic to force that breakthrough into the Social Darwinist fantasy of evolution as progress. Only if we start from the Procrustean assumption that all of evolution is a ladder of breakthroughs leading to us can we cram bats, along with squirrels, foxes, sequoias and other living things into steps on that ladder. For that matter, if bats could write, and suffered from some equivalent of Herbert Spencer's anthropocentrism, they might argue that primates were simply what happened to those relatives of the ancestors of bats that fell behind in the evolutionary struggle toward flight, and had to content themselves with life in the trees.[4]

Like other groups of living things, human communities face pressures from their environments, and adapt or perish in response. Thus it makes sense to speak of human social evolution. Here again, though, the evolutionary process moves outward in all directions rather than ascending through an imaginary hierarchy of levels. Our species started off as hunters and gatherers, but as new opportunities opened up, communities moved into different niches the way the bug-filled night sky attracted the ancestral bats. Where large herbivores could be tamed, nomadic herding societies came into being; where food plants could be raised, tribal agricultural societies were born; where fields of seed-bearing grasses offered the best option for survival, agrarian societies took shape. Just as the placement of extra skin in arboreal animals opened or closed unexpected doors, one of these choices had potentials no one could have foreseen; unlike livestock or garden truck, the main products of herding and tribal agricultural ecologies, grains yielded surpluses large enough to support cities, advanced technology, and the exploitation of fossil fuels that gave the modern world its three centuries of exuberance.

Just as bats are no more evolved than squirrels, industrialism is no more evolved than other human ecologies, and the new human ecologies that emerge in the aftermath of our age will not be more evolved than industrialism; they will simply be different. The breakthrough that gave humans access to fossil fuel created the first technic society and enabled it to outcompete other systems in the same way that an invasive exotic outcompetes less robust native organisms. As fossil fuels deplete and environments change, in turn, succession comes into play, pushing future technic societies toward more sustainable, K-selected forms, while population cycles of boom and bust take place as well.

All this is ecology in action, following patterns found throughout nature across the time scales of population, succession and evolution. On the broadest of those scales, the one evolutionary biologist Stephen Jay Gould named "deep time,"[5] an evolutionary leap unquestionably plays a role in shaping our future, *but that leap has already happened*. The emergence of the first technic societies three centuries ago was a precise equivalent to the leap that sent the first bats winging clumsily through the Paleocene skies, and close equivalents to the potent environmental pressures toward adaptation that shaped those early bats are already coming to bear on humanity. On a narrower time scale, the succession process made inevitable by the looming breakdown of today's R-selected human ecologies is under way, and on a narrower scale still, the fall of one human civilization — ours — is tracing out patterns of population decline familiar from history.

The utopian fantasies that have been piled onto the idea of evolution have no place in these patterns. Human social evolution is likely enough to take unexpected turns in the centuries ahead of us, but nothing justifies the claim that some such turn will bring an end to poverty, war, injustice and environmental destruction. These things did not end when industrialism rose out of agrarian societies, after all, nor did the rise of agrarian, horticultural and nomadic societies bring an end to the social divisions, tribal warfare, and occasional ecological disasters that beset hunting and gathering societies.[6] Evolution is a process of adjustment to circumstances, not a ticket to Utopia; to expect something different from it this time around is to bet the future on the fantasy that a miracle will bail humanity out from the consequences of its own mistakes.

It is worth noting, though, that belief systems founded on the hope of salvation through miracle are common in times of severe cultural crisis. Sociologists, who call them "revitalization movements," have tracked them all along the spectrum of human ecologies from hunter-gatherer bands up through industrial nations.[7] At the core of every revitalization movement is a faith that the troubles of the present will suddenly vanish, and a world of bright promise just as suddenly appear, if the faithful only devote themselves intensely enough to moral, ritual or cultural purity. Such hopes drove such classic revitalization movements as the Ghost Dance of the Native American peoples of the Great Plains; it is one of history's tragedies that it brought them, not to the Utopia the Paiute prophet Wovoka predicted, but rather to the massacre at Wounded Knee.[8]

The decline of civilizations has proven to be a fruitful breeding ground for revitalization movements many times before, and it is no surprise that the same thing is happening again. Nor is it startling that a civilization that has given science much the same role that other cultures assign to religion, as the core foundation for its worldview and the basis for claims to cultural authority, should build at least one of its revitalization movements on concepts borrowed from the sciences. Still, this implies that today's expectations of an evolutionary leap that will solve the world's problems are simply a sign of the times, revealing the intensity of the stress bearing down on people in the industrial world just now. A useful approach to the future ahead must start by setting such fantasies aside and accepting that the currents of change shaping today's world — no matter how unwelcome their impact may be — will not be interrupted by miracles.

the end of affluence

One advantage we have in gauging the possibilities of the immediate future is that this is not the first wave of crisis to hit industrial civilization, but rather the second. The first filled the four decades between the beginning of the First World War in 1914 and the fall of French Indochina in 1954. That time of crisis and disintegration is still close enough to living memory that its phases, rather than the period as a whole, have names we recognize: the First World War, the Russian Revolution and the civil war that followed it, the rise to power of European fascism starting with Mussolini's March on Rome in 1922, the bubble economy of the 1920s, the global Great Depression of the Thirties, the Second World War, and after it, the implosion of the last European colonial empires and the birth of nearly half of today's independent nations.

These events drew much of their momentum from one of the most sweeping changes in modern history: the end of European dominance over the rest of the planet. In 1914, most of the Earth's land surface was either ruled from a European capital or was controlled by nations founded by European immigrants.[9] By 1954, that state of affairs was no longer true. The unraveling of European empires sparked economic crises, political horrors and global power struggles, including two of history's most brutal wars. The era now dawning is likely to see changes on the same scale, and the world that emerges as the current wave of crises ends may be as unfamiliar to us as the world of the 1950s was to those who remembered the 20th century's first decade.

The changes ahead will be shaped by the results of that earlier transformation. The last European empire — the Russian empire in eastern Europe and central Asia that called itself the Soviet Union — came apart in the late 1980s. The last empire ruled by a nation of European origin — the even larger overseas empire of the United States — is arguably headed for the same fate.[10] The details will be different but the cause is the same: the costs of empire sooner or later outrun the benefits, leading to one form or another of national bankruptcy and the implosion of the imperial system.

The first result of the approaching crisis will thus be the twilight of the much-ballyhooed global market of the 20th century's last decades. Globalism was a temporary artifact of a world in which energy costs had been forced so low and economic disparities between nations raised so high that distance didn't matter and arbitraging labor costs across continents seemed to make economic sense. As energy costs climbed in recent years, nations with energy resources have recognized the political side of resource exchange. Those who denounce this as "resource nationalism" seem to have forgotten that the government of Russia, for example, was not elected by Americans, and gains no benefit from policies that benefit American consumers or politicians while disadvantaging their Russian equivalents.

Volatile food prices have pushed this same transformation into overdrive. Governments around the world that once made their ability to feed their people a sacrosanct element of national policy and were talked out of this sensible strategy during the heyday of cheap energy have suddenly discovered that the lukewarm gratitude of foreign politicians and the plaudits of economists in ivory towers count for very little when a hungry mob heads for the presidential palace. During the recent price spikes, most of the Asian countries that produce rice thus banned rice exports so their own people would get enough to eat.

The "free market," for that matter, was never that free in the first place. A slanted playing board designed to maximize the flow of wealth to the world's industrial nations and minimize flows in the other direction, it replaced more straightforward forms of colonialism while maintaining unequal patterns of exchange that allow the five percent of the world's population who live in the United States to use about a third of the world's natural resources.[11] It's not surprising that countries assigned the short end of the stick by these arrangements would throw them off as soon as they could, and the economic instabilities of recent decades gave them an opportunity to do so.

A response from the centers of empire was also inevitable, and it emerged in the late 1990s in the form of the neoconservative movement. There was never much new about the neoconservatives, and even less that was genuinely conservative, but like most of the intellectual fads that bedevil American politics, the neoconservatives had a plan. Though no one was rude enough to mention this in public, that plan was a response to the imminence of peak oil, and found an audience once it became clear that the free market was not going to find a replacement for America's dwindling oil reserves. The plan, outlined by the neoconservative Project for a New American Century, centered on American military occupation of the oil-rich nations of the Middle East, beginning with Iraq, under the threadbare rhetoric of "spreading democracy."[12]

History will not be kind to the neoconservatives. Their plan was poorly conceived and ineptly carried out, its goals are now utterly out of reach and its unintended results may yet include the implosion of American economic and military hegemony. Whether it succeeds in bringing about that final irony, the neoconservative consensus that unites both major parties in the US and its close allies is falling to pieces. The attraction of that consensus was simply that no one else had a plan that would let the United States cling to its precarious position as the world's dominant power. In the wake of the neoconservative debacle, the American political debate has begun to shift slowly from the maintenance of empire to the raw necessities of national survival. It is anyone's guess how long that shift will take, and how much of the world's remaining fossil fuels will be wasted in an attempt to maintain a failing status quo.

Nor is it clear which nations will take advantage of the collapse of American empire to extend their own hegemony. Parts of the world peripheral to today's industrial core will follow their own trajectories. The Muslim world and T'ang dynasty China reached the zenith of their own cultural arcs while the Western world was scraping the bottom of the last dark ages. In the same way, the great cultural and technical developments of the next century could well emerge in what are now impoverished parts of the world. The global reach of industrial civilization, though, makes it unlikely that any nation will escape the approaching troubles entirely, and the equally global drawdown of resources erases the possibility that societies of the future will be able to duplicate the industrial model as it now exists.

One thing that can be taken for granted, though, is that as America's empire winds down, the great majority of Americans will have to get used to living on much less than they, their parents and their grandparents came to expect. The end of American empire — coming on the heels of the peak of conventional petroleum production — marks the end of the age of affluence and the coming of the stage I have called "scarcity industrialism."

the age of scarcity industrialism

This form of human ecology has few modern precedents outside of wartime and the transition to it is likely to see a great many false starts and futile attempts to impose the thinking of the past on the realities of the future. Still, it's not an impossible transition, and will likely be easier than others we'll face along the way. The nature of the transition is straightforward enough. The modern industrial world is geared to constant expansion: of goods and services, technology, energy use, resource extraction and population, among other things. That expansion will no longer continue as the limits to growth begin to bite in the next few years, and many things — including the economic framework of the industrial world — will change accordingly.

As the peak of conventional oil production recedes in the rearview mirror, we approach the beginning of serious declines in energy availability. How serious those will be is a matter for guesswork today, but balancing failing production from existing fields against new production from fields under development and unconventional sources such as tar sands and biodiesel, something on the order of a four to five percent decline per year seems likely for the first decade or so. That will be a body blow to existing economic and social arrangements. Still, production increases of four to five percent a year didn't bring Utopia, and production declines on the same scale won't bring Armageddon, either.

A very large percentage of the energy used in a modern industrial society, after all, is wasted. During an age of cheap abundant energy, it's profitable to use energy in ways that have no real economic value, because the profit to be made selling the energy outweighs the short-term costs of wasting it. Tourism, the world's largest industry just now and a classic example of this logic, uses titanic flows of energy and resources to ship people around the world so they can waste time and spend money. Shut down the tourist industry — as nearly every country in the world did in the Second World War — and put its resources to more productive use, and industrial societies could weather a steep drop in energy supplies without impacting essential goods and services.

The sheer amount of energy that is wasted in a contemporary industrial society provides room for dramatic economies once conservation stops being a slogan and turns into a precondition for collective survival. Transportation, the largest single use for fossil fuels in the industrial world today, is a case in point. Many of today's social habits have evolved, in effect, to maximize the amount of energy that has to be used for transportation; consider, as one example out of many, the land use policies that require homes, shopping areas and workplaces to be so far apart from one another, and so widely dispersed, that neither walking nor public transit is a viable choice for many people. Before the age of cheap petroleum, this was not the case; most people lived within easy walking distance of the places where they worked and shopped. To this day, in cities and towns built before the Second World War, it's far from uncommon to find buildings that once contained groceries and other small businesses on corner lots in the middle of residential districts.

A reorganization of urban space along these older lines will be all but inevitable with the end of the cheap abundant energy that makes today's commuter lifestyles viable. The successful neighborhoods of the age of scarcity industrialism will spring up where jobs can be found, and most such places will be where resources can be had — with the end of cheap transportation, remember, it will no longer make sense to put any more distance between raw materials, processing facilities and consumers than absolutely necessary. Nor will it be functional to scatter homes across acreage that could be more productively used for the truck gardens, chicken farms and other small-scale agricultural establishments that ringed nearly all cities in the industrial world three-quarters of a century ago.

Instead, apartments and row houses will be the order of the day. Where these don't currently exist, materials scavenged from today's less sustainable building projects will be used to fill in the gaps between existing structures; ground floor apartments, converted homes, and the subdivided shells of today's sprawling retail establishments will be filled by local shops and small businesses. The best of these neighborhoods may resemble cityscapes in the poorer regions of contemporary Europe, while the worst may resemble nothing so much as today's Asian and Latin American slum districts: crowded, cluttered and none too clean, they will be far from today's idea of desirable housing, but they will be successful adaptations to the radically changed conditions of the near future.

As this suggests, the fading of the economy of waste promises to stand most of the economic slogans of recent decades on their head. When transportation accounts for most of the cost of commercial products, that fact alone will write RIP on the headstone of the global economy, because goods made overseas will be priced out of markets dominated by local production and regional trading networks. The cutting edge of the new resource nationalism is already visible, as energy resources and strategic materials become the mainsprings of power and governments start treating them accordingly. Expect this to expand dramatically in the decades to come, as dependence on foreign resources becomes a noose around a nation's neck, and economic independence — even at a lower standard of living — becomes the key to national survival.

The pendulum of power has already begun to swing away from the multinational corporations that have had so much influence in recent years toward national governments willing to use force to maintain their territorial integrity and control over resources. The recent failure of attempts by the world's rich nations to impose trade treaties on the rest of the planet is one marker of this shift. When most resource transfers across borders are negotiated between governments, rather than purchased on the open market, those whose power comes solely from money will have much less clout than they have today. Those governments that master the new calculus of power soonest, in turn, will dominate the age of scarcity industrialism.

However it unfolds, though, the age of scarcity industrialism will be no more permanent than the age of abundance industrialism that preceded it. While it lasts, control of fossil fuels and other nonrenewable resources will be the key to power and prosperity, but by that very token fossil fuels and other nonrenewable resources will be exploited and depleted. As resource production in one nation after another drops too low to support an industrial system, industrial economies will unravel and give way to other forms of human ecology — or, in the terms used earlier in this book, other seral stages in the succession that leads to the ecotechnic societies of the future.

What cannot be predicted in advance is which nations will survive the transition to scarcity industrialism and which will crack under the strain. The United States could go either way. The nations that claw their way to the top of the heap under one set of economic conditions rarely hold on to the same status when conditions change and America's fervent commitment to the economics of waste has sent fissures of weakness throughout its structure; the implosion of America's empire is thus a foregone conclusion. If the next generation of American politicians are lucky and smart, the US might be able to coast down the curve of declining empire as Britain has. If not, the US could face any of the unpleasant destinies that await failing empires, ranging from decades of economic and cultural stagnation to nightmare scenarios of political and military collapse, followed by partition by hostile powers. Hard as it may be to imagine such possibilities in our future, history reminds us that such things happen, and nothing justifies the claim that they cannot happen again.

the age of salvage

The reality of natural limits provides one of the core themes of the succession process facing industrial societies. In the age of scarcity industrialism, the power of limits will dominate political discourse; as that age gives way to societies based on salvaged resources, the power of limits will affect every aspect of human life. At that point familiar social and economic patterns come into play.

Archeologists around the world recognize the distinctive traces of a collapsed society, and one of these is the recycling of old structures for new uses. In the ruins of the old Mayan city of Tikal, excavations have unearthed traces of the people who lived there after the Maya collapse. In this quiet afterword to the city's history, the palaces of the lords of Tikal became the homes of a little community of farmers and hunters who scratched out a living in the ruins of the city and made their cooking fires and their simple pottery in the midst of crumbling splendor. The same thing appears in dead civilizations around the globe. The logic behind it, though, has not often been recognized: when a civilization breaks down, the most efficient economies are most often those that use its legacies as raw material.

To understand how this works, it's necessary to detour a bit and discuss H.T. Odum's useful concept of *emergy*, or *em*bodied en*ergy*.[13] Emergy is the total energy needed to produce a good or provide a service, including all the energy and material that went into making the good or service available. A coffee cup sitting next to your computer, for example, embodies the energy needed to mine and process the clay, provide raw materials for the glaze, fire the kiln and ship both the raw materials to the factory and the finished cup to you. That amount of energy is the emergy cost of the cup: without that much energy, you can't have that cup — or at least you can't get the cup in the same way.

When energy sources are cheap and abundant, emergy basically doesn't matter. The lords of Tikal didn't have to worry about the effort their work crews devoted to hauling, carving and setting up stone stelae any more than their equivalents today have to worry about the fuel that ships coffee cups, and the coffee that fills them, halfway around the planet. On the downslope of collapse, on the other hand, emergy matters a great deal, and the most abundant source of readily accessible emergy consists of the material remains of the collapsed civilization. To the surviving people of Tikal, it was much more efficient to use the crumbling palaces of a bygone age for shelter and concentrate their limited resources on the hard work of making a living in a damaged environment than it would have been to build homes somewhere away from the ruined city.

The fantastic amounts of energy flung around casually by industrial societies today will make this an even more viable strategy, once the fossil fuels that make industrial civilization possible go the way of Tikal's glory. Steel, the most widely used metal nowadays, is the best example. A fifty-foot steel girder in a skyscraper contains a great deal of emergy, because the ore has to be mined, smelted, purified, cast, milled, shipped thousands of miles and lifted far above the street, using huge amounts of energy, before it finally settles into place in a new building. Since most of the iron ore mined today is low-grade taconite containing less than five percent iron, and nearly all of today's iron and steel production uses energy-intensive methods to turn out high-quality steel, that girder contains far more emergy than a salvage society can use on anything but essentials.

In order to put the emergy in that girder to use in a salvage society, on the other hand, a few simple tools and the energy input of ordinary human labor will be enough. A hacksaw can cut the girder free and chop it into workable parts, a wagon pulled by horses or oxen can haul the metal and a blacksmith's hammer, anvil and charcoal-burning forge can pound it into nails, knives, plows, saws and a thousand other useful things. The economics of metalworking in a nonindustrial society make this an attractive proposition, since one fifty-foot girder of structural steel will keep a village blacksmith supplied with high-quality material for many years of normal production.

Now it's true that the same village blacksmith could smelt his own raw material from bog iron—the iron sulfide laid down in most temperate zone wetlands by chemosynthetic bacteria. There is plenty of bog iron, since it has not been mined commercially in centuries, and most North American deposits away from the Atlantic coast have never been mined at all. Bog iron can easily be smelted into workable form—people in Dark Age Europe and early colonial America did it with heat from charcoal—and the same thing can be done as easily with rust, which is iron oxide, the most commonly used iron ore in the days before fossil fuel subsidies made it possible to use low-grade ores such as taconite.

Still, the steel unintentionally stocked up for the future by today's civilization makes a far better source. A small proportion of that stockpile consists of high-temperature alloys that can only be worked with modern technology, but the vast majority — girders, pipes, auto frames, sheet steel and much more — can be forged at temperatures much lower than the ones needed for smelting ore and yield better metal into the bargain. They will be the primary metal source in the age of salvage, and there are billions of tons of steel in every habitable corner of today's industrial world, enough to keep salvage societies supplied for a very long time.

At the same time, steel is only one of hundreds of raw materials free for scavenging from the ruins of today's cities and towns. The value of copper and aluminum in houses has risen high enough already that the unsold subdivisions thrown up in the late housing bubble are now being stripped of their copper wiring and aluminum window frames by thieves who sell the resulting metal at a tidy price. What counts as a criminal enterprise today will likely become a growth industry in the deindustrial future.

Nor are raw materials the only legacies of the industrial age the salvage societies of the future will use. A good deal depends on how much technical knowledge survives the cycles of crisis before the age of salvage begins. Communities that keep electrical generators in working order, for example, can use salvaged equipment that runs on electricity. Internal combustion engines may still be viable on a small scale if they can be fueled with ethanol or biodiesel. In a deindustrializing world, access to such technologies will be a potent source of economic, political and military power, and this guarantees that they will be used.

Like the age of scarcity industrialism before it, though, the age of salvage will be self-limiting, because the economics that support it also guarantee the exhaustion of the resources that make it possible. Eventually, no matter how many times they are patched and rebuilt, the last of the Old Time machines will stop running. There will be no more overgrown storage centers and long-abandoned suburbs to mine for valuables, and in time even the vast ruined cities of the ancients will yield up the last of their metal, though the seral stage of salvage will have effectively wound down well before then. Once salvage no longer forms the primary resource for the economies of the future, the seedtime of the ecotechnic age will begin.

the coming of the ecotechnic age

The three-phase model of future history I've suggested here is easy to misunderstand. The seres outlined above are ideal types — they are the kind of things we can expect, rather than anything more exact — and the forms these patterns take on the ground of actual history will be complex, messy and idiosyncratic. This should come as no surprise, since the same thing can be said of the phases commonly used to sort out history that has already happened.

When a historian says that England embraced a mercantilist economic system in the 16th century, for instance, he does not mean that the English economy shifted gears all at once on January 1, 1501, or any date thereafter. Nor does he mean that the English economy of that century lacked features of the older feudal economy or foreshadowings of the capitalist economy that replaced mercantilism later on, nor that the English mercantilist economy was identical to all other mercantilist systems. Rather, he means that the traits implied by the term "mercantilism" — an export-based economy geared toward generating a favorable balance of trade with competing nations, foreign policy initiatives pursuing overseas colonies and the expansion of sea power and the like — provide a workable sketch of the shape toward which the English economy moved during the century in question.

The same rule applies to the model of the future sketched here. The transition from today's economy of abundance to the scarcity industrialism of the near future, for example, will likely be just as slow and ragged as the rise of mercantilism. Some nations — Russia comes to mind — have already implemented the political control of resources that forms a core feature of the coming phase, while other nations have barely begun to move in that direction. For that matter, the 1950s-era American autos cruising down the streets of Havana, repeatedly rebuilt with jerry-built parts, show that core features of the salvage economy are already present in some parts of the world right now.

Thus the world of fifty years from now will include nations at many different points along the sequence of phases. It will likely be dominated by nations that have successfully managed the transition to scarcity industrialism, while the powers of today's age of abundance will be the fallen empires and failed states of that time. Meanwhile, those nations that drew short straws in the geopolitical lottery and have no fossil fuels, may already be well into the salvage society phase, mining the legacies of the industrial age to meet local needs and pay for whatever foreign trade can still be had. Nations that lack both fossil fuels and salvage, in turn, will either return to agrarian, nomadic or hunter-gatherer economies or, given luck and a good foundation in ecology, may be pioneering the first rough sketches of an ecotechnic society.

Fast forward another century, when Hubbert's curve has bottomed out and fossil fuels will be rare geological specimens, and the powers of the age of scarcity industrialism will likely have collapsed in turn. Those areas with plenty of salvageable scrap and the political and military savvy to hold onto them will be the regional powers of a world in which global reach no longer exists, while other areas — the modern conception of the nation-state will likely be abandoned by then — will have only sustainable resources to rely on; some of those will have settled into some nonindustrial mode for the long term, while others may be building an ecotechnic system.

All these changes will take place amid the turmoil of ordinary history, that unending and uneven rhythm of crises and the rise and fall of governments and peoples that tends to hide broader shifts from contemporary eyes. The short-term effects of changes in climate and environment can also hide the slower transformation of succession, the movement toward K-selected sustainability unfolding in the wake of the evolutionary leap that gave rise to the world's first technic societies three centuries ago. All three time scales discussed earlier — the rise and fall of civilizations, the successional shift from R-selected to K-selected human ecologies and the vaster sweep of evolution itself — must be taken into account.

The generations that grow up in a world after industrialism will face many of the same kind of challenges that their ancestors did in the dark ages that followed other high civilizations. Some of those challenges must be confronted as they emerge. It may well be possible, however, to counter or even forestall others by drawing on the resources of industrial civilization, to hand down valuable tools and insights to those who will need them. While Utopia is not an option, societies that are humane, cultured and sustainable are quite another matter. There have been plenty of them in the past; there can be many more in the future; and actions we can take today can help make that goal more accessible to the people of the ecotechnic age. The practical possibilities open to actions of this sort will be the theme of the next eight chapters of this book.

PART II

RESOURCES

Preparations 5

THE PARALLELS BETWEEN biological evolution and the transformations of human societies are crucial to an ecological vision of the future, but the comparison breaks down in one place. As far as anyone knows, other species cannot foresee changes in their environments; they can only react to those changes as they appear. In theory, at least, human beings are not quite so limited. Our ability to anticipate the future is far from limitless, but the attempt can be made.

It is crucial in this context to get past the common assumption that social changes only happen because people intentionally choose them, and are limited only by people's willingness to make them happen. Like so many of today's blind spots, this has its roots in the illusion of independence discussed in Chapter One. Not that long ago, by contrast, most social thinkers placed human societies in a wider context in which limits and tradeoffs were as appropriate as they were necessary. This more organic approach was central to the thought of social philosophers such as Edmund Burke and was enshrined in the elegant balances of the American constitution, but it finds few listeners in contemporary society.

These days even Karl Marx's time-release Utopia has been shouldered aside to make room for new ideologies of the Left, promising a more immediate delivery of the perfect society. Meanwhile the Right has by and large abandoned its role as the conserver of social practices — this is what the word "conservative" means, after all — and seeks to impose its own utopian visions on a recalcitrant world. Pick an ideology, any ideology, as close to the mainstream or as far out on the fringe as you like, and you can expect to hear its proponents announcing that a perfect world can be had for the asking, if only people accept their particular recipe for making one.

The contrast between the bright future their scheme will supposedly bring and the horrors that the world will suffer otherwise is a constant feature in this sort of rhetoric. Interestingly, though, nothing at all need connect the threatening crisis to the proposed solution. Not long ago, for example, I read a lively essay arguing that the best way to bring humanity into harmony with the environment was for the world to embrace socialism. Leave aside the fact that this is as likely just now as a resumption of the Crimean War; the more important point is that it doesn't solve the problem it claims to address. On the theoretical level, shifting ownership of the modes of production does not affect how those modes interact with the ecosystem. In terms of history, socialist countries have had as bad a record when it comes to the environment as capitalist countries. Instead of finding a solution to the problem, in other words, the author simply found a new way to promote his preferred agenda.

The rhetoric of survivalism rests on the same dubious reasoning. Survivalists identify a problem, any problem, insist that it will inevitably lead to the collapse of civilization into a Road Warrior future populated with rampaging mobs convenient for target practice, and present the survivalist answer as the only possible response. The ritual incantations of politicians seeking office present the same thinking in an even more caricatured form: no matter what the problem happens to be, the solution always amounts to throwing out the last scoundrel who got into office promising to solve it so another scoundrel can take a swing at it. As the limits to growth find their way back into common discourse in the years ahead of us, every project for social change you care to imagine will likely try to redefine itself as the answer the world is waiting for.

projecting the shadow

This is not always simply a rhetorical tactic. The belief in an imminent evolutionary leap critiqued in Chapter Four is only one version of a dizzying array of contemporary political and quasi-political movements that believe devoutly that the future is about to drop into their hands. Fascists hoping to replay their triumphs of the 1930s, Marxists dreaming of proletarian revolution, social justice activists yearning for a redistribution of wealth, Klansmen waiting for the South to rise again, neoprimitivists hoping that civilization will go away and take six billion inconvenient lives with it so they can lead the hunting and gathering lifestyle of their dreams, and many others pin their hopes for solving the world's problems on the imminent end of the industrial age. Not since Doctor Fox's Genuine Arkansas Snake Oil stopped being sold on the carnival circuit has one remedy been applied to so many different diagnoses.

It takes only a few minutes on an online forum or a meeting anywhere on the political fringes to find people who believe the next step in history will be the birth of whatever sort of society they most want to inhabit. Some of them come to the discussion with detailed plans for the perfect future, backed up figuratively — and now and again literally — with a backpack full of citations cherry-picked from favorite authors and the media. On the other end of the spectrum are those who have no idea what the world of the future will look like but cling to an unshakable faith that it will have to be better than the world of today.

A good friend of mine once recounted a conversation he had in the last days of 1999 with someone who confessed she was deeply worried about the Y2K problem. He assumed that she meant she was worried about the struggle for survival in the aftermath of the massive systems collapse some people were still predicting at that point, but she quickly set him right. Her job was unsatisfying, her marriage was on the rocks and her life was at a standstill. What worried her was that she might wake up on January 1, 2000 and find that nothing had changed.

It might seem that a boring job, a troubled marriage and a midlife crisis would be preferable to starving to death in a burned-out basement in the aftermath of cataclysmic social collapse. My friend's story is far from unique, though, and the fact that so many people today treat catastrophe as an inkblot onto which to project their fantasies of a better life is one of the most troubling signs of our times. To some extent, this follows from the common habit of imposing one's daydreams onto an unknown future; it's always part of the narrative of apocalypse that dieoff only happens to other people. No matter how unready they are for the strenuous task of surviving the collapse of a civilization, every would-be survivor seems to expect a place among the lucky few. Still, there are deeper patterns at work here.

The collapse of the New Left in the wake of the Sixties and the abandonment of traditional conservatism by the ideological Right well before the Reagan era left a political vacuum not yet filled. For some years now, as a result, most radicals have pictured their task in the purely reactive language of resistance and opposition, while the mainstream parties abandoned their old commitments in favor of the pursuit of business as usual for its own sake. This has spared all sides the daunting challenge of coming up with constructive proposals for the future. The downside, though, is that those who sense the necessity for change are left with nothing but fantasies of a perfect world to feed their hopes. These encourage people to forget that every other collapse of civilization in history led not to Utopia, but to a harrowing age of warfare, migration, population decline, impoverishment and the loss of priceless cultural treasures. Just as revolutionaries who insist that nothing can be worse than the status quo are often surprised to find just how much worse things can get, those who insist that today's industrial societies are the worst of all possible worlds may find themselves pining for the good old days of suburbs and freeways if they get the collapse they think they want.

Furthermore, the last few decades have birthed a culture of political demonology, especially but not only in America, in which the slight differences between competing political parties have been redefined in terms of absolute good and evil. Vigorous debate over the merits of candidates for office is the lifeblood of a republic, but when opponents of a public official are unable to walk past his portrait on a wall without screaming obscenities at it — and I have seen this on both sides of the political chasm in America today — something has gone very wrong.

Carl Jung's concept of "projecting the shadow" is relevant here.[1] Too many Americans have fallen into the seductive but disastrous habit of blaming their political adversaries for their own feelings of shame and resentment. Even the briefest glance at history shows where that sort of scapegoat logic leads and it's no place any sane human being would want to go. A good deal of what happened during Germany's ordeal between 1933 and 1945, as Jung pointed out in a prescient essay,[2] can best be understood as the end result of this sort of projection, with a grand Wagnerian *Götterdammerung* as finale. It's entirely possible that some similar madness could grip America in the years to come.

Whether or not anything so ghastly happens, the unfolding crisis of industrial society is likely to bring in a bumper crop of self-defeating ideologies, unless we can stop thinking of modern industrial society as either the best or the worst of human cultures and recognize that it contains much that is worth saving alongside much that belongs in history's dumpster. As the age of cheap abundant energy comes to an end, many things will change, and for most people those changes will bring more challenges and fewer comforts, a surplus of troubles and shortages of almost everything else. New social forms will inevitably come into being, but their positive features will be tempered by the sharp realities of resource limits, social chaos and a biosphere in disarray.

This future begins from the hard fact that industrial civilization is on the downslope of its history, headed for the scrap heap alongside the dead civilizations of the past. The crises now taking shape around us are simply the current manifestations of patterns that shaped the fall of other civilizations. As I have suggested in *The Long Descent*, those patterns will most likely play out in a long decline ending in a dark age in which much of the technology, knowledge and cultural heritage of today is at risk of being lost. The approaches discussed in this book are meant to help bring the most useful legacies of the present through the successional process to the far future, and help catalyze the rise of ecotechnic societies at the far end of that process. Even so, the ecotechnic age will be no more utopian than today's world. It may well have moved past some of today's problems, but it will surely face problems of its own. The habit of projecting the shadow tends to foster beliefs that anything less than perfection is unacceptable, but this sort of thinking is a luxury only affordable in ages of abundance.

the effects of homeostasis

The dream of apocalypse as a door to Utopia has also encouraged the unproductive habit, very common in today's activist circles, of identifying any available crisis as the final blow that will bring about the sudden unraveling of the existing order and its replacement by a better world. It's understandable that passionate believers in the arrival of some new system would see the first signs of its appearance everywhere, but when that habit leads to a string of false predictions, a little realism might be helpful.

Nearly all these inaccurate predictions are rooted in a failure to grasp a central idea of ecology, the principle of *homeostasis*. One of the crucial ways in which any living system — a cell, an organism, an ecosystem or a planet — differs from nonliving matter is that living systems respond to change with efforts to maintain balance. The little fence lizards that show up on summer days in my garden show homeostasis in action: in the morning, when the ground is cold, the lizards scamper over to the nearest patch of sun to warm up; by the time the day has become warm, the lizards are cooling off in the shade. This is the basic structure of homeostasis: negative feedback — countering cold with heat or heat with cold, or any other extreme with its opposite — works to keep a system within the range of conditions it needs for survival. It does not always work, for the lizards or for anything else, but its effects must always be taken into account.

Human societies, like fence lizards, are organic systems, and they respond to changes in their environments in much the same way. This is the factor that most predictions of sudden collapse miss: when faced with crisis, societies take action in response. Competent leadership can help this process along, but it is not necessary; the last century or so has been so rich in crises that many homeostatic processes function automatically. Thus when petroleum prices spiked upwards to $143 a barrel, oil companies intensified the search for new oil fields and reopened wells that were uneconomical in the halcyon days of cheap oil. When petroleum prices dropped to $35 a barrel, in turn, oil companies shut down unproductive wells and delayed investment in new oil fields. These changes helped dampen out price swings, and the price swings themselves were generated by another homeostatic process that balanced the supply of oil with the demand.

The upward end of these swings imposed, in effect, a form of rationing by price on petroleum supply, in which those who have money get oil and those who lack it go begging. This is a deeply unfair way to distribute a limited resource; unless the industrial world goes through drastic political changes in the near future, though, it's what will be in place as shortages begin. Rationing by price also has one practical advantage: the ration coupons (we call them "money") and the infrastructure for using them are already in place, and most people have some access to both. Steps to make rationing by price less unfair are certainly viable, and the possibility of establishing some more equitable rationing system down the road is certainly worth exploring. All of these are homeostatic processes on another level, meant to manage declining fuel availability in a way that does as little damage as possible to the fabric of communities and societies.

Other homeostatic feedback loops can be traced in recent headlines. In Oregon, where I lived until recently, much of the long and fertile Willamette Valley has been used for years to grow grass seed for the lawn-improvement market. When the housing bubble imploded, the market for grass seed plunged, while the market for wheat has been boosted by shortages. As a result, farmers up and down the Willamette are planting wheat instead of grass. Other agricultural regions that have concentrated on specialty products are finding good financial reasons to return to staple food crops; once again, this is homeostasis at work.

These adjustments have their grimmer side, of course. As food prices go up, the poor and disenfranchised worldwide are sliding closer to the edge of starvation. That is a tragedy, and a moral crisis of no small magnitude. Like so many of the unwelcome changes approaching us, though, it is nothing new in human history. As recently as the 19th century, famines racked most nations regularly, and a good half of the population even in industrial nations lived in poverty. Most people in the industrial world nowadays have forgotten how much deprivation was a part of ordinary life before the brief 20th-century heyday of cheap abundant energy.

As fossil fuels become scarcer and more of the global food supply is diverted to biofuel production, increased starvation in the world's poorer countries and a sharp increase in the world's roster of failed states are among the likely results. Meanwhile, economic, political and cultural readjustments will hit the industrial world as income redistributes itself from urban centers to farm country. These shifts will be harder to track as speculation and economic volatility send prices of many agricultural commodities soaring upwards and crashing back to earth in vertiginous cycles of boom and bust. Behind the volatility, though, the general trend will be upwards; many centuries will likely pass before food is ever again as cheap compared to incomes as it was in the second half of the 20th century.

All this is homeostasis on another scale, part of the process by which the massive growth of human populations made temporarily possible by cheap abundant energy is being balanced out by negative feedback imposed by the limits of the biosphere. It is easy to think of such things in apocalyptic terms; it is much harder to realize that similar levels of famine, violence and social turmoil have been a normal part of human experience for most of our species' history, and are matters of living memory for many people today. Nor are human social structures as fragile and incapable of meaningful response to severe crisis as some modern ideologies assume.

Counterintuitive as it sounds, the tendency for social systems to remain fixed in place becomes more effective, not less, in the face of sudden catastrophe. Consider the Black Death in medieval Europe.[3] As an example of mass dieoff, it's hard to beat — the first terrible epidemic of 1346–1351 killed a third of the population of Europe, and recurring outbreaks that came every decade or so took up to ten percent of the survivors each time. In the form of peasant revolts, it even managed to produce a semblance of the marauding hordes that play so large a part in survivalist fantasies today. Still, despite the horrific death rate, social disorder and huge cultural impacts of the Black Death, European civilization did not collapse. The survivors simply picked themselves up and went on with things much as before.

Nearly all the sudden catastrophes imagined today would turn out the same way. If sudden mass dieoff swept over America today, after all, the survivors would still have the material culture of pre-collapse society as a resource base. Libraries, schools and businesses would still exist; abandoned buildings and machines could be put back to use; viable energy sources such as hydroelectric dams could be restored to working order; it might even be possible to reactivate the Internet once power started flowing again. Everyone alive after the collapse would have grown up in the pre-collapse world and learned the skills needed to operate a modern society. Before long, something like today's culture would have reestablished itself, just as late medieval cultures across Europe reestablished themselves after the Black Death.

A complete recovery is only possible, though, if there is a short time between collapse and aftermath. Consider an extended decline and a much less promising picture emerges. First, and most obviously, a gradual decline takes much longer. By the time the process is finished, the people who remember how an advanced civilization used to function are in their graves and anything perishable has long since perished. It's one thing to break into an abandoned library two years after a sudden collapse, when most of the books will be dusty but readable; it's another thing to do the same thing two hundred years after the beginning of decline, when the books have crumbled into sawdust because they were printed on high-acid paper or rotted after the rains got in. In the same way, the Internet might well be able to withstand any number of sudden crises, but the death of a thousand cuts inflicted by inadequate funding, maintenance failures, spare parts shortages, power failures and the struggle for increasingly scarce resources is quite another matter.

the twilight of technology

In the twilight of other civilizations, furthermore, the long curve of decline was punctuated by periods of crisis in which urgent needs absorbed all available resources for decades. During these periods, anything not relevant to the needs of the moment will go begging for maintenance and upkeep if it isn't stripped for spare parts, burned as heating fuel or destroyed in war, rioting or the like. This is the other side of homeostasis: the efforts of societies to find balance in difficult times, while they can help survivors of sudden catastrophe to recover, can amplify the effects of a slow decline. Each crisis becomes a bottleneck through which only part of a society's culture and knowledge can pass; repeat the process often enough and very little remains.

Technology, to which so many people nowadays assign the role of savior, can turn into a double-edged sword at such times. When people talk about the role of technology in the future, most of the time the technologies they have in mind are the flashy ones — those that have not been around long enough to slip into the background texture of everyday existence. Especially in periods of decline, though, the fate of societies is often determined by the survival or disappearance of technologies so common they are hardly noticed.

For the people of Easter Island, for example, deepwater canoes had been part of daily life for thousands of years before their ancestors reached this westernmost outpost of Polynesia. This may be among the core reasons that nobody on Easter Island anticipated the consequences of cutting down too many trees. The resulting deforestation eliminated large tree trunks, an essential resource without which deepwater canoes could not be made, cutting off the majority of the island's food supply and, at the same time, the only way out of the trap the Easter Islanders set for themselves. The canoe had been so omnipresent a part of life for so long that the possibility of its absence very likely never entered into the islanders' darkest dreams.[4]

A similar sort of dependence on a vulnerable technology, according to the medieval Arab historian ibn Khaldûn, drove the collapse and abandonment of cities across the Middle East and North Africa in the centuries prior to his own time. The *Muqaddimah*, ibn Khaldûn's treatise on the forces that shape history, paid close attention to the relationship between settled agricultural civilizations and nomadic herding societies.[5] The relationship is worth watching. As far back as ancient Sumer, which in historical terms is as far back as you can go, the ebb and flow of power between desert herdspeople and settled agriculturalists set the heartbeat of history. Civilizations rose on the backs of new technologies, prospered and expanded at the expense of their nomadic neighbors, transmitted their technical skills to those same neighbors, and then faltered and collapsed beneath nomad incursions.

What sets ibn Khaldûn apart from the other historians who tracked this cycle is his attention to the role of technology in bringing the cycle to a disastrous end. From Sumerian times onward, irrigation canals formed the backbone of settled life across the Middle

East. While desert irrigation can cause salinization (the slow buildup of salts in the soil), this does not follow as automatically or as disastrously as some current writers insist — it is too rarely noticed that large amounts of wheat and barley are still grown in most of the nations of the Fertile Crescent. What made the agrarian economies of the desert nations vulnerable, rather, was the same failure to prioritize the requirements of essential technologies that doomed the Easter Islanders.

By the Middle Ages, due to climate change and topsoil loss, most Middle Eastern societies had to direct much of their economic output into maintaining the canals and waterworks on which survival depended. This became their Achilles' heel, because the desert nomads who conquered the urban centers rarely grasped the importance of the irrigation systems and starved them of resources until local breakdowns could no longer be repaired and the entire system failed. Like the deforestation of Easter Island, the breakdown of the irrigation canals was a one-way ticket to collapse; once farmland turned into desert, the agricultural wealth that made canal building and repair possible was no longer there to be spent, and regions that had been settled for millennia turned into deserts spotted with crumbling ruins.

All this has uncomfortable echoes in the present. Like the inhabitants of Easter Island, our societies depend on the reckless exploitation of limited resources; like the civilizations of the Middle East whose fate was chronicled by ibn Khaldûn, our survival depends on fragile infrastructures that few of us understand and most of our leaders seem willing to starve of necessary resources for the sake of short-term political advantage. Further, the industrial system that supports us has been in place long enough that most of us seem to be unable to conceive of circumstances in which it might no longer be there.

The first stage of the ensuing crisis has already arrived. What large trees were to the Easter Islanders and irrigation canals were to the medieval Middle East, the money economy is to industrial society, and the speculative delusions that passed for financial innovation over the last few decades have played much the same role as the invading nomads of ibn Khaldûn's history. The result, just as in the 1930s, is that a nation still relatively rich in resources and a large and skilled labor force is sliding into crushing poverty because the intricate social system we use to allocate labor and resources has broken down.

Those who think of catastrophe as the only way industrial civilization can fall are unlikely to notice the gradual shifts that lead to the same destination at a slower pace. It's as though the boy who cried wolf, in the famous parable, was convinced that an immense army of wolves the size of horses would suddenly swoop down and eat up all the sheep in the world at once, and mistook every whistle of wind in the trees for the distant howling of the wolf pack to end all wolf packs. Meanwhile, real wolves — scruffy, undersized and depressingly few in number compared to the massed uber-wolves of the fantasy — were picking off a sheep or two each day from the fringes of the flock.

It may seem obvious that if you're guarding sheep from wolves, and wolf packs numbering fewer than fifty are beneath your notice, your sheep are going to be eaten. Still, similar points have gone unnoticed by a surprising number of people these days. Thus it's crucial to gauge the shape of the future and the possible responses to it on the basis of local conditions and the specific resources that can be brought to bear on them, and to remember that the declining years of other civilizations in the past can offer valuable insights into the ways that collapse plays out in the real world.

adaptive responses

Instead of the planned futures and apocalyptic transformations imagined by too many people nowadays, a more effective response to the future will need to take into account the homeostatic processes that give stability to current social patterns, and the need to approach the future one step at a time. The future that comes into being through such a response, in turn, will neither be made to order according to some deliberate plan or imposed on humanity by sudden radical changes. Rather, it will emerge through a process of negotiation in which humanity's efforts and nature's responses both play a role. In a word, it will be adaptive.

This is a rich word with a double meaning. In the jargon of evolutionary biology, "adaptive" is used to label anything that allows an organism to respond effectively to the demands of its environment. When the environment is stable, what makes an organism adaptive stays the same from generation to generation. When the environment changes, though, what is adaptive changes as well. Genetic variations that would have been problematic under old conditions become advantages under the new; if the shift is large enough, a new species emerges. This points up the other, dictionary definition of the word — according to my Webster's Ninth, "showing or having a capacity for or tendency toward adaptation."

Both these meanings have crucial roles in the work ahead as industrial society breaks down. On the one hand, it's crucial to find ways of living that are adaptive in the ecological sense — that is, well suited to the new reality of scarce energy and hard environmental limits. At the same time, we will not be landing plump in that new reality overnight, nor do we know in advance what that new reality will look like, so it's just as crucial to find ways of living that are adaptive in the dictionary sense — that is, capable of adapting to the unpredictable changes of the transition.

Some basic guidelines for adaptive approaches can be sketched out here. First, an adaptive response is *scalable* — it can be started and tested on a small scale, with modest resources, and scaled up from there if it proves successful. Second, an adaptive response is *resilient* — it remains useful under changing conditions and can respond creatively to pressures. Third, an adaptive response is *modular* — it can be separated into distinct elements, which can be replaced at different scales and technological levels. Finally, an adaptive response is *open* — it does not demand assent to any particular ideology or belief system but rather works within many different ways of thought and life.

These perspectives follow the lead of one of the most thoughtful and least remembered works from the appropriate technology movement of the 1970s, Warren Johnson's *Muddling Toward Frugality*. Johnson argues that the end of fossil-fueled affluence is a given, and trying to fight it makes about as much sense as playing King Canute and trying to order back the incoming tide. Rather, he suggests, we need to live with it—and the best way to do so is to take the modest, piecemeal, unimpressive steps that will get us through the crises of the future.

One of the things that makes *Muddling Toward Frugality* so valuable a guide is that Johnson deals directly with the cultural narratives underlying projects for social change. The habit of relying on ideologies and grand plans, he suggests, is borrowed from the language of tragedy, in which great heroes risk everything for an ideal. This makes great literature and drama, of course. Still, the heroes of tragedy usually die and often take everything they care about down with them, so they may not be the best model for constructive change!

As an alternative, Johnson offers the unexpected possibility of the comic hero. Comic heroes are usually muddlers, stumbling cluelessly through situations with no grander agenda than coming out the other side with a whole skin. They are the opposite of heroic and their efforts at muddling through fail to inspire the kind of reverence so many proponents of social change seem to long for. Unlike tragic heroes, though, they usually do come out alive on the other side of the story and they often bring the rest of the cast with them.

The decline and fall of modern industrial civilization may not seem like promising material for comedy, but the basic strategy of muddling has much to recommend it. As mentioned before, no one knows what an ecotechnic civilization would look like. No one knows what steps will be needed to make the transition from an industrial society to an ecotechnic one. No one knows how fast fossil fuel production will decline, how the resulting economic shock-waves will play out, how soon the effects of global climate change will begin to impact today's societies in a big way or any other answers to the crucial questions we face. Nor does anyone know what responses will deal with these things effectively, if any of them do.

What we do know is that certain things are not working now and need to be changed; and that certain other things that still work may not work for long, so that having replacements handy is probably wise. In a situation of this sort, betting everything on one grandiose plan for the future is a poor bet. A wiser approach would encourage many different responses to the situation. In the same way, piece-meal responses that focus on narrowly defined dimensions of the crisis and can be implemented on a small scale before moving to a larger one may be more useful than grand attempts to change everything all at once.

Thus embracing an adaptive approach to the situation, and then simply trying to adapt, may get us further than any plan at all. *Solvitur ambulando* is an old Latin phrase that still gets a little literary use in English these days. Taken literally, it means, "it is solved by walking." A more idiomatic translation might be, "you'll find the answer as you go." An adaptive approach to the crisis of industrial society could well take this as a watchword. Yet such an approach requires a sharply different way of thinking about solutions than most people in the peak oil community have pursued so far.

a time for dissensus

This has not always been grasped, even by those who have a good sense of the crisis ahead of us. Back in the 1970s, when the limits to growth first showed up on the radar screens of the modern world's collective discourse, a great many contributors to that dialogue insisted that global planning, and only global planning, could back humanity away from the brink. Since then, plans with a similar agenda have appeared at regular intervals. The durable environmentalist Lester Brown is one example. He released the original version of his

Plan B in 2003; version 3.0 appeared while this book was being written. More versions will no doubt be forthcoming.

If the crisis we face could be met by making plans, however, we would have little to worry about. The archives of most American city and county governments contain energy plans drafted and adopted, after extensive citizen input, in the 1970s, calling for conservation standards, public transit projects, zoning changes to reduce the need for cars and many other sensible things. Nearly all these plans have been gathering dust since the Reagan years and the rest were shoved aside in the 1990s to make way for the recent housing bubble. Skepticism toward another round of planning may thus be in order.

Those who try to plan an ecotechnic society today are in the position of a hapless engineer tasked in 1947 with drafting a plan to produce software for computers that did not exist yet. At that time, nobody knew whether digital or analog computers were the wave of the future. The experimental computers that existed then used relays, mechanical linkages, vacuum tubes and other outmoded technologies, while most of the components that made it possible to build today's computers had not yet been invented or even imagined. Under those conditions, the only plan worth making would consist of one sentence: "Invest heavily in basic research and see what you can do with the results." Anything else would have been wasted breath, and the more detailed the plan, the more useless it would have been. The lesson here is a crucial one: meaningful planning can only take place when the outlines of the solution are already known.

Many of today's students of the future nonetheless start from the assumption that an effective response to the challenges facing industrial society requires reaching consensus around some plan of action and then carrying it out. The raw uncertainties of the future ahead of us, though, make this a dubious proposition. Even if some common denominator of agreement could be found among competing views of the future, it could at most cover a small fraction of the possible options, and there is no way of knowing whether those particular options are in fact the best possibilities to explore. A quest for consensus thus risks narrowing the options at a time when the range of potential choices needs to be as broad as possible.

This is where a different approach, *dissensus*, comes into play. Dissensus, a concept coined by postmodern theorist Ewa Ziarek, is the deliberate avoidance of consensus.[6] It has its place when consensus, for one reason or another, is either impossible or unwise: for instance, when irreducible differences make it impossible to find common ground for agreement on the points that matter, or when settling on any common decision would be premature. In situations of these kinds, encouraging people to pursue conflicting and even diametrically opposed options increases the chance that someone will happen on an answer that works.

Dissensus is not simply the vacuum left by a lack of consensus. It has its own methods, values and style, and its results differ in kind from those of consensus or other methods of decision-making such as majority rule. The most creative periods in the arts are generally times of dissensus; it is precisely when innovative minds reject the consensus or the majority opinion of their time and strike out along the lines of their own individual inspirations that the most innovative cultural creations come into being. Nearly all great artists are masters of dissensus, and so is the greatest artist of all, Nature.

The first inch-long vertebrates who darted about in shallow seas half a billion years ago, for example, did not come to some sort of genetic consensus about where evolution was going to take them, nor did the evolutionary process itself push them in one direction. Some of their offspring became fish, some amphibians, some reptiles, some birds and some mammals; one of the latter wrote this book, and some others are reading it. Evolution is dissensus in action: the outward pressure of genetic diversification running up against environmental limits and, now and then, pushing through to some new adaptation, such as the wings of bats, the opposable thumbs of primates or the cultural evolution of human beings.

As we enter a future of harsh limits and unpredictable opportunities, this is arguably the kind of organic process we need most. For this reason, the chapters that follow will not attempt to set out rules for how an ecotechnic society should function, much less for how one ought to be built. Instead, I will sketch out some of the ideas, tools and techniques available today that prefigure the approaches that an ecotechnic society will need to develop or that are likely to prove helpful in the nearer term — or, where possible, that accomplish both these necessary goals at the same time. Neither necessity can be neglected in the complex transitions ahead of us.

the mariner's two hands

In the face of so vast a challenge as the collapse of a civilization, no plan is foolproof. Even with the help of history, it is quite possible to misjudge the future disastrously, and even those who guess right may not be able to avoid the future's dangers. This is why the concept of dissensus is so vital just now. In the face of unpredictable change, increasing the range of variation in a species makes it more likely that some members of the species will have what it takes to adapt. Rather than trying to find points of agreement among those seeking to respond to the crisis of industrial society, then, it may be wiser to encourage disagreement and even competition, in order to further the widest possible range of experimental ventures.

The need for dissensus does not simply apply to the technologies different individuals and groups might decide to pursue, the organizations they might choose to launch or support or the survival strategies that might seem most promising to them. It also reaches into the realm of ends. One of the central challenges of our predicament is that no one today knows what kinds of human society will be well suited to a world after industrialism. The more diverse the visions we imagine here and now, the more likely it will be that some of the actions taken in pursuit of those visions will stumble across viable roads to the future.

The difference between those measures that will help in the short term and those valuable later on also needs to be kept in mind. Here again, it is important to avoid the fallacy of assuming that it's possible to leap from today's society to an ecotechnic world in a single massive transformation. The obsession with sudden catastrophes mentioned earlier in this chapter plays a role in making this fallacy seem reasonable, and so does the natural tendency of those who believe they have imagined a better society to try to build it here and now. The lessons taught by succession show why this so often fails catastrophically: the best society imaginable cannot survive if the conditions in which it must exist do not support it or favor some other society more.

This means that a constructive response to the future must include two sets of strategies. First, the challenges of the immediate future need to be faced, using whatever resources are best suited to the needs of the moment, whether or not those resources will be there when the ecotechnic age begins. Second, those things that will likely be of use when the seres of scarcity industrialism and salvage are over need to be preserved wherever possible, so that our descendants will have the legacies they need from our time as they begin to craft the cultures and technologies that will rise from the ruins of the industrial world.

An old rule from the days of tall ships told the mariner that when he was up in the rigging, one of his hands was for himself, and the other was for the ship. He was within his rights to cling like grim death to the nearest spar or ratline with one hand, so the other would be free to haul lines and handle sails. The same principle could usefully guide those who take up the challenge of preparing for a difficult and uncertain future. One set of skills, resources and tools will deal with the immediate impacts of the crisis of industrial civilization, while another set will preserve the useful legacies of the modern world for generations to come. Human nature being what it is, the first half of that equation is likely to get a great deal more attention than the second, but both have their place.

The proposals in the chapters that follow are made with these points in mind. They are illustrations of the types of things that might prove useful in getting through the first waves of change facing the world's industrial societies, and beginning to lay foundations for the ecotechnic age of the far future. None of them are guaranteed to work. Those who find themselves disagreeing with one or another point made below, or even with all of them, are well advised to consider their disagreements carefully and then pursue their own insights rather than mine. Dissensus plays no favorites; the ideas suggested in this book, or any book, are only a small part of the range of possibilities that deserve to be explored as the long trek to the ecotechnic future begins.

Food 6

One of the great gifts of crisis is that it points out what is essential and what is not. As we descend along the curve of deindustrialization, that gift is likely to arrive in horse doctor's doses. Those who insist that the top priority in an age of faltering petroleum production ought to be finding some way to fuel a suburban lifestyle or to preserve some exotic technology — the Internet, say, or space travel — risk finding out the hard way that other things come first.

At the top of the list of those other things are the immediate necessities of human life: breathable air, drinkable water, edible food. Lacking those, nothing else matters. The first two are provided by natural cycles that industrial civilization has done its best to disrupt, but so far the damage has been localized and attempts by business interests to control the world's water supply — troubling though these are — are already splintering as the transition to scarcity industrialism hands power back to national and regional governments. There are still crucial issues to consider and hard work to be done, but the sheer resilience of a billion-year-old biosphere that has shrugged off ice ages and asteroid impacts gives humanity a powerful ally.

Food is another matter. Nearly all the food we eat today comes from human-managed systems rather than the free operation of natural cycles, and human population has passed so far beyond the limits of natural ecosystems that trying to feed even a fraction of one percent of today's humanity on wild foods would be a recipe for global ecological catastrophe. Today the agricultural systems that produce our food depend on nonrenewable resources such as fossil fuels and mineral phosphates, and all these are being used up rapidly. Another set of resources that are renewable over long timescales, such as topsoil and "fossil water" from dryland aquifers last filled during the Ice Age, are also being used up rapidly enough to guarantee serious supply problems in the not too distant future, and even renewable resources that are replaced over short time scales, such as water from lakes and rivers, are being exploited so extravagantly that shortages are a real risk. Some peak oil theorists have worried publicly on this basis that declining petroleum production could cause the sudden collapse of industrial agriculture, followed by worldwide starvation. If today's agriculture were to follow its present course, this might be a serious risk.

Here as elsewhere, though, homeostatic processes make straight-line extrapolations the least likely guide to the future. Today's industrial agriculture did not pop fully formed out of a John Deere plant like Athena from the head of Zeus. It evolved gradually as farmers and businesses in the 20th century took advantage of the ready availability of fossil fuels. At the time, putting more energy into agriculture was adaptive; it made use of a cheap abundant resource to make up for other resources that were more expensive or less available. That same equation, though, works just as well the other way. As fossil fuels and agricultural chemicals become more expensive, farmers will use less of them and replace them with other resources.

Thus the transition from industrial agriculture to ecotechnic agriculture need not happen in a single leap. Nor could it do so. The transition from an R-selected agriculture to a K-selected one requires more than replacing chemical fertilizers with compost, diesel tractor fuel with biodiesel and so on; it needs massive changes in everything from eating habits and land use policy to food distribution networks and waste management. Those changes will take time; like the broader transition from industrial society to the ecotechnic future, they will pass through intermediate stages along the way.

the next agriculture

The first phase of the agricultural transition is already taking place. The organic farming revolution, the most important aspect of that transformation, may be the most promising and least discussed of the factors shaping the future. It's not a small factor, either: in recent years conversion of farmland from conventional to organic production has been taking place at the rate of several million acres a year.[1]

Because it needs no chemical fertilizers or pesticides, organic agriculture uses less fossil fuels than standard agriculture, and it produces equivalent yields. Ecological benefits aside, it has a simple economic advantage: since he does not have to pay for agricultural chemicals, an organic farmer can raise crops more cheaply than his conventional neighbor. Organic produce generally commands better prices as well. As raw materials for the chemical industry become scarce, that equation will become even harder to ignore. In the meantime the infrastructure and knowledge base necessary for organic farming on a commercial scale is already in place and continues to expand.

The forms of organic farming used in most commercial operations mark a major improvement over conventional farming, but more richly sustainable methods are already waiting in the wings. Biointensive organic gardening, pioneered in the 1970s by John Jeavons in California and John Seymour in England, cycles compost and organic matter into deep-worked garden beds to produce very high yields on small plots — high enough that using these methods, it's possible to feed one person a spare but adequate diet year round on 1,000 square feet of soil.[2] Similar in its potential is the biodynamic method invented in the early 20th century by Austrian mystic Rudolf Steiner, using complex organic supplements to build soil fertility.[3]

More ambitious still is permaculture, developed in the last quarter of the 20th century by Australians Bill Mollison and David Holmgren.[4] The system's name, a contraction of the phrase "permanent agriculture," describes its central practice — the construction of artificial ecosystems, primarily of trees and perennial plants, that produce food and raw materials for human beings. It has been slow to catch on among commercial farmers, partly because it requires more training and knowledge than other approaches, partly because it relies on unusual forest crops with limited markets, and partly because its annual yields per acre are lower than most of the alternatives. Still, the potentials of the method are intriguing enough to attract a community of enthusiasts, who are continuing to develop the method and explore its practical applications.

These are only the most widely discussed of the alternative methods of food production evolving within the shell of today's agriculture. Many others exist; a thorough survey of the new growing methods could fill a book by itself. Nearly all of them make minimal use of fossil fuels, either as energy or as raw materials for agricultural compounds, and equally slight use of technologies more complex than a hoe or a compost bin. Hydroponics and other technology-heavy methods had their fifteen minutes of fame in the alternative agriculture scene decades back and then dropped out of use as people in that scene realized that more dependence on energy inputs and complex gear was the opposite of what was needed. The focus on simple tools and attentiveness to natural cycles that pervades alternative agriculture these days has encouraged exactly those methods that will be most useful as fossil fuels run short.

The wider network of exchange and transportation that gets food from farm to table is a thornier problem, especially but not only in North America. The world-class railroad system that once connected North American farmland to markets and ports across the continent has suffered from decades of malign neglect and it's an open question whether enough of it can be renovated in the teeth of economic crisis and political decline to make a difference. The canal system that preceded rail over the eastern third of the continent is in even worse shape, though some critical links remain.

The transportation revolution of the 20th century had many consequences that will have to be undone in a hurry, but its impact on the human food chain may be the most extreme. For the first time in history, it became possible to centralize agriculture so drastically that only a tiny fraction of food was grown within a thousand miles of where it was eaten. Traditional habits that relied on local produce gave way to today's reliance on commercially processed prepared foods. All of this made food far more costly in energy terms, but in the heyday of fossil fuels nobody worried about that.

As we move further into the 21st century, though, the industrial food chain of the late 20th century has become a costly, brittle anachronism. Swings in food prices have added economic uncertainty to the daunting challenges of feeding a hungry world. As fossil fuel depletion begins to price industrial agriculture out of the market, other ways of farming are moving up to take its place; the downside is that each of these exacts its price, and all require additional investments in an age when those may well be in short supply. Replace diesel oil with biodiesel and part of your cropland has to go into oilseeds; replace tractors with horses and part of your cropland has to go into feed; convert more farmland into small farms serving local communities and economies of scale go away, leading to rising costs.

This is not the only reason why industrial agriculture is unlikely to go away soon. The corporate agriculture sector remains a massive economic presence, and in a society where money can be cashed in for political influence, the industrial farm lobby will almost certainly keep the playing field slanted in favor of its own model of agriculture. At the same time, this doesn't mean that industrial agriculture will function well. Far more likely is a situation in which soaring fossil fuel prices cascade down the food chain, turning industrial farms and their distribution networks into economic basket cases propped up by government subsidies, sky-high food prices and trade barriers that keep other options out of the economic mainstream.

Since people still need to eat, a future of this sort will likely accelerate the rise of microfarms and market gardens and the cooperatives, farmer's markets and community-supported agriculture schemes that provide food distribution outside the official economy. This backyard agriculture will have to use minimal fossil fuel inputs and rely on local distribution, since fuel costs will put long-distance transport out of reach. It will have to focus on intensive production from very small plots, since most acreage large enough for industrial farming will already be occupied by industrial farms, and thus hard to come by. It will also benefit greatly by relying on human labor with hand tools, since the economic effects of peak oil will likely send unemployment rates soaring while making machinery and fuel challenging to provide. The next generation of organic growing methods fill all of these requirements and already has a substantial following among home gardeners.

Attempts to replace shrinking petroleum production with ethanol and biodiesel add fuel to the same blaze. With today's agricultural methods, ethanol from corn yields negative net energy, but this will likely change as farmers learn to use less fossil fuels and industrial processes improve. At the same time, such schemes will likely have a serious impact on food prices and, ultimately, food availability. Large industrial farms already monopolize the market for ethanol feedstock, and can be counted on to continue to dominate that sector of the farm economy. As industrial agriculture shifts from growing food to producing fuel, backyard agriculture will become one of the few options available to fill the void in the food supply. In the process, the economics of small-scale farms and market gardens will improve even further.

As the transition to scarcity industrialism proceeds, North America could easily end up with something like the agricultural system the former Soviet Union had in its last years, with vast industrial farms that had become almost irrelevant to the national food supply and an underground economy of backyard market gardens that produced most of the food people actually ate.[5] That divided system could last as long as scarcity industrialism does, but it's unlikely to last much longer. Industrial agriculture was a creation of the petroleum age: steam-powered tractors fired by coal were tried at several points in the 19th century, but horses remained a better bargain until the advent of the diesel tractor in the middle decades of the 20th. The transportation network that delivers food to consumers, of course, is wholly dependent on petroleum derived fuels.

As that network breaks down, changes in the agricultural system will ripple upwards through the rest of society, forcing unexpected adjustments in areas of the economy that have no obvious link with agriculture at all. Rising prices and unsteady production will pinch budgets, impact public health and make malnutrition a significant issue all through the developed world. Actual famines are likely, as shifting climate interacts with an agricultural economy in the throes of change. All this is part of the price of adaptation as the agriculture of abundance industrialism gives way to systems better suited to the needs of the deindustrial future.

When petroleum production has declined to the point that tractors are no longer viable, and local production and distribution costs much less than centralized production and distribution through a fraying transportation net, the big industrial farms will go out of business. Thereafter, the future will belong to food producing methods that work best within the specific constraints of local ecosystems. Out of those changes — which may take centuries to work themselves out — the agriculture of the ecotechnic age will gradually take shape.

compost as template

One process central to most alternative growing methods, the homely art of composting, deserves special attention here. As a fertilizer source and a method of waste disposal, it will likely be important all through the deindustrial transition. By turning waste products into resources, furthermore, it transforms a linear process into a circular one, and so fulfills one of the essential requirements of a K-selected human ecology.

There is nothing new about composting; this is one of the most elegant things about it.[6] Nature composts relentlessly. Dig a shallow trench through the topsoil of a meadow, and you can watch organic matter being turned into soil by a labor force of earthworms, bugs, fungi, microbes and plant roots. The art of composting consists of setting up the conditions to put this natural process into overdrive. More than a century of research and practical experience in composting has yielded methods that work well on any scale from household to community.

These possibilities are matched by the sheer potential for composting. I have been unable to find even an approximate figure for the total volume of compostable waste generated annually in the United States, or any other industrial country for that matter. It's certainly a huge figure, and the amount that goes into landfills rather than being composted into fertile soil is not much smaller.

Perhaps the best way to make sense of the process and its implications is to follow composting through its cycle on the small scale, in a household setting. For practical reasons, I will use the example I know best.

My wife and I produce between one and four cubic feet of compostable waste a week, depending mostly on the season of the year, the produce available at the local farmer's market and the state of our garden. All of it goes into a compost bin of black recycled plastic in the back yard. So does another half cubic foot per week from a friend's kitchen; his living situation doesn't allow him to have a bin of his own, so he contributes to ours. All the peelings and scraps from our kitchen and his go into the compost bin, along with garden weeds, plants that have passed their season and other yard and garden waste, leavened with double handfuls of dried leaves saved from last autumn. Those are the only inputs, other than a little labor with a shovel once a month or so to keep the pile working. The pile compacts dramatically as it works, and once a year the hatch at the bottom of the compost bin disgorges the output — eight to ten cubic feet of black, damp, sweet-smelling compost, ready to be worked into our garden beds.

This output is potent stuff. The exact mix of nutrients in compost depends on the raw materials that go into it, but an ordinary blend of kitchen and garden scraps will normally yield compost rich enough that soil amended with compost rarely needs any other fertilizer at all. At least as important as mineral nutrients such as phosphorus and potassium are humic acids and other complex organic substances, which make minerals more available to plants, increase the soil's capacity to hold water and feed earthworms and other beneficial organisms.

The composting process is also self-sterilizing, because decomposition generates heat — 150° to 160°F is a fairly common temperature for the core of a good compost pile. Most weeds, seeds and pathogens die quickly at composting temperatures; this is why compost should be turned at intervals while it works, so everything in the bin can pass through the core and be cooked. The heat produced by composting is significant enough that bins in warm climates need to be placed out of the summer sun and may have to be hosed down on hot days; compost heaps have been known to burst into flame when the heat of decomposition rose past the ignition temperature of the pile's more flammable ingredients.

Compost can be made on an industrial scale — some businesses and public utilities do this — but the resulting product cannot simply be used as a replacement for chemical fertilizers. It requires different farming methods and, for best results, a different philosophy of agriculture, oriented toward long-term sustainability rather than short-term gains. Thus composting is not an effective way to maintain business as usual, but rather a bridge beyond the industrial age to the ecotechnic future. It has the four characteristics of an adaptive response discussed in Chapter Five. It is scalable — composting systems have worked well on every scale from small households to the organic waste collection systems of entire cities. It is resilient — it has wide margins for error and can adapt to a broad range of conditions. It is modular — each step in the composting process can be done in many different ways. And it is open — it requires no ideological commitments beyond an unwillingness to waste soil nutrients and can be practiced in almost any social context that can be imagined.

Nearly every difference between composting and the industrial way of dealing with waste points up a significant difference between an R-selected and a K-selected human ecology. First and most obviously, where industrial civilization turns resources into waste, composting turns waste into resources. Today's industrial agriculture, like the rest of our R-selected technic society, is a linear process that turns fossil fuels and mineral deposits into pollution, and fertile soil, clean water and breathable air into waste management problems. Composting bends the line into a circle, turning waste into a resource that can be used to meet crucial agricultural needs.

Second, where industrial civilization works against natural processes, composting works with them. The industrial world's concept of progress sees it as the conquest of nature, and this way of thinking leads industrial societies to see natural processes as obstacles to be overcome or enemies to be crushed. The result is a self-defeating misuse of resources in which farming methods convert soil into a nearly sterile mineral medium which must then be soaked with chemical fertilizers to make up for the lost fertility that natural cycles create in healthy soil. The composting approach feeds those cycles instead of disrupting them and thus accomplishes the same goal with less effort and expense.

Third, where industrial civilization requires complex, delicate and expensive technologies to function, composting — because it relies on natural processes that have evolved over countless millions of years — thrives on a much simpler and sturdier technological basis. Set the factory complexes, energy inputs and resource flows needed to manufacture NPK (nitrogen-phosphorus-potassium) fertilizer side by side with the simple bin and shovel needed to produce compost and the difference is hard to miss. If your small town or urban neighborhood had to build and provide energy and raw materials for one or the other from scratch, using the resources available locally right now, the difference would be even more noticeable.

Fourth, where industrial civilization is centralized and can only function on geographic scales large enough to make centralized infrastructure profitable, composting is decentralized and functions best on a small scale. Among the reasons why no small town or urban neighborhood will ever build a factory to produce NPK fertilizer is the simple fact that the cost of the factory equipment, energy supply and raw materials would be far greater than the return. A backyard fertilizer factory for every home would be even more absurd, but a backyard compost bin for every home is arguably the most efficient way to put composting technology to use.

Fifth, where industrial civilization degrades its environment, composting improves its environment. On a finite planet, the more of a nonrenewable resource you extract, the more energy and raw materials are needed to extract what is left; the more of a persistent pollutant you dump into the environment, the more energy and raw materials are needed to keep the pollutant from interfering with other activities. Thus industrial civilization climbs a steepening slope of its own making until it finally falls off and crashes back to earth. By contrast, the closed loop that runs from composting bin to garden plot to kitchen and back to composting bin becomes more effective, not less, as the cycle turns: for example, rising nutrient levels in the garden plot lead to increased yields at harvest, and thus to increased input to the compost bin.

Finally, all these factors mean that where industrial civilization is brittle, composting is durable. It's a common lesson of history and ecology alike that the intricate arrangements made possible by periods of stability shred like cobwebs in a gale once stability breaks down and the environment lurches toward a new equilibrium. In a time of turbulence, systems that depend on concentrated resources, intricate technologies and massive economies of scale face a much higher risk of collapse than systems that lack these vulnerabilities.

Many other sustainable technologies embrace one or more of these same factors. So far, not many embrace all of them. Even technologies as promising as metal recycling have a long way to go before they become as close to the ecotechnic ideal as composting. Comparisons of this sort point up the way that sustainable techniques such as composting, beyond their practical value, can be used as templates for a much wider range of approaches.

These refinements are not limited to the realm of the technical. The contrast between the monumental absurdity of industrial society's linear transformation of resource to waste, on the one hand, and the elegant cycle of resource to resource manifested in the humble compost bin on the other, makes it hard to avoid challenging questions about the nature of human existence, the shape of history, the meaning of the cycles of life and death and the relationship of humanity to the source of its existence, however that may be defined. Over the centuries to come, perspectives such as these are likely to shape the collective conversations of the societies that succeed ours.

in the dark with both hands

The need for composting and similar resource cycling methods in the deindustrializing future is almost impossible to exaggerate. Most of the agricultural chemicals used to produce food today are made from nonrenewable resources, and require a great deal of energy to mine and process. As energy supplies dwindle, using increased amounts of energy to extract minerals from dwindling deposits quickly becomes a losing bet. This equation promises to upend current arrangements across the industrial economy, but its impact on agriculture may be the most severe of all.

After generations of industrial agriculture, much of the world's arable land is so depleted that it can no longer produce crops without significant inputs of nitrogen, potassium, phosphorus and other minerals. Unless the deindustrializing world can find a readily available, abundant and concentrated source of plant nutrients to replace chemical fertilizers before fossil fuels begin to run short, it's not much of an exaggeration to say that future humanity may face a Hobson's choice, in which dwindling supplies of fossil fuels can be used to produce fertilizer to keep people from starving to death, or electricity and heat to keep them from freezing in the dark — but not both.

Fortunately a readily available, abundant and concentrated source of plant nutrients has already been identified by researchers. The technologies needed to obtain and process the raw material into high-grade fertilizer are thoroughly tested — they have been used for thousands of years in many parts of the world — and require no nonrenewable resources at all. The only reason this valuable resource is not already being exploited on a massive scale all over the world is that most people nowadays seem unable to distinguish its source from a hole in the ground.

The material in question, of course, is human fecal matter — or, as one book on the subject has usefully relabeled it, "humanure."[7] The average human being in the industrial world produces between 2.5 and 3 pounds of feces a day, or approximately half a ton of raw fertilizer feedstock per year. Multiply this by the 300 million residents of the United States, and then factor in the equally massive waste streams generated by domestic animals and livestock, only a part of which is currently used for fertilizer, and you have some sense of the scale of the resource that we are quite literally flushing down the toilet.

The technology that converts this resource into fertilizer most effectively is the same art of composting just discussed. Composting turns feces into a concentrated, odorless fertilizer that resembles nothing so much as ordinary soil. Nor does this product pose health risks; a well-maintained compost pile is a fiercely Darwinian place in which organisms bred in the sheltered setting of the human colon do not last long. Many studies have shown that fecal matter, after it has been competently composted, contains no more human pathogens than ordinary soil.

The other side of the human waste stream, urine, is another significant agricultural resource that is currently piddled away. The average person in the industrial world produces around a third of a gallon of urine a day, or a little over a hundred gallons a year of liquid fertilizer feedstock rich in nitrogen, phosphorus and potassium — the three minerals most needed to improve the fertility of depleted soil. Urine from healthy human beings is sterile, it can be converted into fertilizer using any of several simple, proven methods, and it could play a major role in transitioning away from NPK fertilizers produced from nonrenewable sources.[8]

So why has the world been unable to get its fertilizer together on this issue? What keeps composted humanure and urine from being a primary resource base for farmers struggling to replace dwindling inorganic sources of plant nutrients? Much of the reason reaches deep into the crawl spaces of the industrial world's imagination. People who object to composting human waste very often cite concerns about pathogens or odors, but it rarely takes long to reach the emotional level of a five-year-old clenching his eyes shut and squealing, "Ewww, ick!"

C.S. Lewis pointed out many years ago in a thoughtful study, *The Abolition of Man*, and with much greater force in his science-fantasy novel *That Hideous Strength*, that modern attitudes about dirt and biological waste have their source in what might be called biophobia — a pathological fear of the realities of biological life, coupled with an obsessive fascination with the sterile, the mechanical and the lifeless. Biophobia guides the creation of human environments so sterile that, according to recent research, many currently widespread illnesses may be caused by excessive cleanliness.[9] The same attitude, I'm convinced, drives the horror many people feel when faced with the prospect of eating food fertilized with composted humanure.

Biophobia is one of many modern attitudes that will not survive the transition from an R-selected human ecology of linear flows to a K-selected ecotechnic world of sustainable cycles. Today's economists take the tangled exchanges, multiple roles and mixed motives of real market economies and sterilize them into neat flowcharts that move matter from suppliers to producers and distributors and then to consumers before vanishing into thin air. Food systems built on the same pattern take nutrients from natural deposits, put them into soil, haul the resulting crops into a baroque system of manufacturing and distribution before they get to people and then dump the resulting waste into the world's fresh water supply. This is not a sustainable approach by any definition, and will inevitably change as cheap energy goes away.

It's a measure of how pervasive biophobic attitudes have become that following nature's patterns and cycling "waste" back around to become a resource seems so unthinkable. Still, the possibilities of humanure compost and urine soil amendments offer a glimpse at the agriculture of the future — if we can get our heads out of our fertilizer supply long enough to notice.

pieces of the puzzle

Composting, with or without humanure, is only one way of closing the circle and cycling organic wastes back into soil. Another technique that has received much attention in certain branches of the alternative agriculture movement is sheet mulching: spreading a thin layer of uncomposted organic material over the top of the soil, and leaving it there to rot.[10] This keeps moisture in the soil, stops many weeds from sprouting and cycles organic matter back into humus to improve soil fertility. In dryland bioregions, in particular, it's a key technique for organic growers.

Still, mulching is not a panacea, and it is more effective in some regions than in others. In the part of Oregon where I used to live, for example, slugs are serious garden pests, and sheet mulch is a slug magnet; if you mulch early in the growing season, slugs will eat most of your spring crops. Many organic gardeners there use sheet mulching to overwinter the garden, from harvest's end to planting time, and then dig the mulch under when preparing garden beds for new crops. Many of these same gardeners compost kitchen scraps and yard waste, and so organic material enters the soil by both routes. Different materials follow distinct trajectories: carrot peels go into the compost bin while autumn leaves become sheet mulching, then rot into humus once they're turned under in spring. The two methods do not conflict with one another, and the same springtime digging that turns the mulch under also works in the year's supply of compost from the bin.

Nor does this exhaust the possibilities open to the ecotechnic agricultures of the future; there are still other options for closing the loop and cycling organic matter back into the soil. One is green manure — this term refers to planting a cover crop of clover or some other nitrogen-fixing plant in the fall, letting it grow all winter, and then turning it under in the spring. Kitchen scraps can also be fed to chickens, rabbits or some other livestock and their manure turned into plant food. A worm bin can replace the usual composting methods, using redworms to break down the organic matter in place of bacteria. A clever idea from the appropriate technology movement of the Seventies can also be used: set up an aquaculture system, feed spare organic matter to tilapia or some other tasty and nourishing fish and use water full of fish feces to irrigate the crops.

Which of these is the answer to the challenge of post-peak food production? Put that way, the advantages of dissensus are obvious, because none of them is *the* answer. All of them, singly or in combination, and other methods not yet invented, can be workable responses to some of the needs people will have as they try to keep themselves and their families fed as our society skids down the far side of Hubbert's peak. Put another way, they are pieces of a puzzle; each has its place, but no one piece completes the puzzle by itself.

This same logic can be applied elsewhere, and offers a clear glimpse of the way dissensus plays out in practice. One common dispute in the peak oil community is whether farmers should continue to use tractors, or whether draft horses will prove to be more viable. Both sides have good arguments. On the one hand, a large farm running tractors on homegrown biodiesel can keep them fueled by devoting ten percent or so of its acreage to oilseed crops, while it takes around thirty percent of acreage to produce fodder for draft horses to provide the same amount of power. On the other hand, you don't need a factory or its substantial inputs of energy and resources to manufacture horses — they do it themselves, with noticeable enthusiasm and no tools other than the ones nature gave them — and a properly fed horse also produces large amounts of excellent organic fertilizer, a significant value that tractors don't provide.

Which is the best option? That depends on a galaxy of factors, few of which can be predicted on the basis of abstract arguments. In those parts of the world that still have working factories and energy resources in the age of scarcity industrialism, tractors may turn out to be more viable; in those regions where the industrial economy survives only in fragmentary form or goes away altogether, draft horses may well have more to offer. Issues of scale, crop and climate are also crucial; the option that would work best for a 160-acre family-run wheat farm on the Great Plains might be disastrous for a 10-acre farm growing vegetables on the outskirts of a West Coast city.

For that matter, neither horses nor tractors have any place in the sort of backyard mixed gardens that had so crucial a role in helping people in the former Soviet Union survive its collapse and may play the same role in getting Americans through a similar collapse in the not-too-distant future. Such gardens require only hand tools and human labor. Intensive gardening and extensive field agriculture are not the same thing, but both will likely have important roles to play in feeding people in the deindustrial era.

The incessant disputes over the value of animal raising in sustainable agriculture yield readily to the same logic.[11] Today's methods of animal husbandry feed livestock with industrially raised grains, adding the sustainability problems of modern grain agriculture to their own serious difficulties. This does not mean, however, that all forms of animal raising must share the same problems. Small livestock — chickens, rabbits and the like — raised on a household scale have provided useful amounts of protein to indigenous peoples around the world for millennia. Ventures along these lines carried out during the 1970s suggest that the same thing could be done sustainably in a deindustrializing world. Larger livestock have long been viable parts of rural agrarian society and their usefulness is unlikely to come to an end as agrarian resettlement becomes a significant trend.

There may well be other possibilities that careful research or blind luck could turn up.

The crucial point to keep in mind is that nobody knows which techniques will turn out to be the best options for either the ecotechnic civilizations of the future or the transitional cultures that will rise and fall on the way there. For the last three centuries all our experience of technic societies has been on the opposite end of the spectrum — those civilizations that burned through resources at the fastest pace they could manage. As a species, we have followed that road just about as far as it can go, far enough that the dead end at its terminus should be visible to anyone who is willing to notice it. All we know for certain is that the way we have been producing food for the last half century has no future; the systems that will replace today's industrial agriculture in the short term are not too hard to imagine, but further out the possibilities become impossible to calculate. Here and now, the best we can probably do is to make sure that as many promising methods reach the middle future as possible, in the hope that those methods that work well in the changing biosphere of that time will find waiting hands to receive them.

Home

RIGHT AFTER FOOD on the list of human necessities comes the patterning of resources in space that ecologists sum up with the word "habitat"—the environmental framework that provides shelter and other necessities of life for any living thing. For human beings in all agrarian and technic societies, and even in some nomadic ones, the most important ingredient in habitat is a home—a built structure that not only provides shelter from the weather, but also plays a central role in framing most human economic and social activities.

Le Corbusier, one of the founders of the International style of architecture, missed the boat when he described a home as "a machine for living."[1] (A home burdened with the rigid purposiveness of a machine would be a very poor place to live, as in fact most homes built according to Le Corbusier's principles have proved to be.) Rather, a home is a *habitat* for living; it could even be called the natural habitat of technic and agrarian humanity. Like other habitats, it should be a stable ecosystem with its own internal energy flows and resource cycles, forming part of a nested series of wider ecological relationships, from local ecosystems through bioregions to the biosphere of the Earth itself. Homes flourish or fail depending on their successful integration with those larger patterns.

The traditional homes of many nonindustrial cultures embody millennia of evolution and provide excellent human habitat, while leaving the wider environmental systems around them relatively unharmed. Most homes in the modern industrial world, from the sprawling McMansions of the well-to-do down to the apartments that house the poor, provide relatively poor habitat for human beings; unlike more traditional housing, which is designed to foster many of the activities of human life, many modern residences make room for sleeping, consuming manufactured products and very little else. They are also hopelessly out of harmony with their surroundings and manage to keep functioning at all only at the cost of massive damage to local and global ecosystems and equally huge inputs of energy and raw materials drawn mostly from fossil fuels.

As the costs of maintaining such dysfunctional human habitats becomes too severe to bear, concentrated energy sources run out and the economy of abundance that created most of today's housing stock comes to an end, homes will have to undergo massive redefinition to deal with the new reality of scarcity. Those that can be modified will need thorough rebuilding; those that cannot will have to be torn down, converted into raw material for salvage and replaced. The detached single-family house will likely become a great deal less common than it is today, and other forms of housing that economize on space and energy may become much more widespread.

Poor energy efficiency is the most obvious problem with today's housing stock but it is far from the only one. Another, even more significant, is the way that so many homes have been designed to exclude all but a small fraction of the activities carried on in traditional human habitats. A home is not just a shelter from the elements and a place to sleep; it is also an economic unit, and in every human culture outside the modern industrial world, it is a *productive* economic unit, not simply a staging area for consumption. There are solid reasons why this is the case; when energy is scarce and expensive, it becomes more economical to produce as many goods and services as possible where they will be used, using labor and resources already present in that place.

Thus the transition toward the ecotechnic future imposes at least two challenges on housing. First, the homes of the future must scale back the energy inputs needed to keep them livable — ultimately, to the point that nearly all the energy needed in the course of ordinary life is produced in or immediately around the home. Second, the homes of the future must once again become economically productive places, where many of the goods and services needed by their residents are produced as well as consumed. These necessities weave together in many ways, some obvious and others surprising, but it will be useful to start exploring them one at a time.

tomorrow's homes

A cliché in alternative circles these days argues that the homes of the future ought to harmonize with nature. Points raised already in this book offer a good deal of support for this claim, but the concept itself needs to be understood with some care. Over the last century or so, terms such as "natural" and "organic" have been caught up in a variety of cultural struggles and come out much the worse for wear. The resulting confusion can be traced all through contemporary culture, but an example worth tracing here is its impact on one of the towering creative minds of the 20th century, American architect Frank Lloyd Wright.

I should confess at the start that I have been a fan of Wright's work for decades, and not merely because he was one of the handful of first-rate talents influenced by the modern Druid tradition.[2] In his quest for an organic architecture, he reshaped the vocabulary of space and form in ways that are still being explored by architects today, and he also produced rather more than his share of stunningly beautiful buildings. Still, there are few geniuses whose works have no flaws, and Wright was not among them.

Stewart Brand ably sets out the case for the prosecution.[3] To begin with, he notes, Wright's roofs leak. Since one essential purpose of shelter is to keep weather out, and making a roof watertight is not that difficult, Wright's problems with this basic task are not a good sign. More broadly, Wright paid more attention to the aesthetics of building materials than their structural qualities, and built a good many splendid buildings that could not hold up to normal wear and tear, or in some cases, the mere force of gravity. Many Wright buildings have had to be torn down due to severe structural problems. Similar difficulties run through every aspect of his work; beautiful as his houses are, for example, they are not always functional or comfortable places for people to live.

Something is arguably wrong with an organic architecture that produces poor habitats for living things. The reason for Wright's difficulties, though, are not hard to trace; like many innovators, he found it easier to outline the territory he hoped to explore than to fill in all the details. His organic architecture was organic in its relationship to the site and the aesthetics of his materials, but it fell short of its goal in at least two other ways. The first, as already mentioned, was the lack of a functional relationship between his designs and the strengths and weaknesses of the materials he used; in a fully organic architecture, the placement of a beam or the slope of a roof ought arguably to depend as much on the need to hold up a floor or keep out the rain as on the aesthetics of site and structure.

The second aspect is subtler, and Brand's book *How Buildings Learn* is perhaps the best guide to it. A building exists in time as well as space, and it grows and changes throughout its life from the first clearing of the site to the last swing of the wrecking ball. Successful buildings adapt to the people who live in them or use them, just as the people adapt to the buildings; Brand argues that this is how buildings "learn." In this sense, many of Wright's buildings were very slow learners, and some proved to be wholly unteachable. Of course a great deal of modern architecture suffers from severe learning disabilities; it does not help that the architect's job, in Wright's time as today, most often ends when the blueprints are handed over to the builder, and the cult of creative genius that pervades modern society encourages innovative architects to ignore the grubby realm of practical realities. Still, an architecture that attempts to be organic in any meaningful sense needs to grapple with these issues and, where possible, overcome them.

The challenge posed by Wright's successes as well as his failures is to do exactly that: to interpret his guiding concept of organic architecture in senses that go beyond the esthetic, and to create successful human habitats that relate harmoniously with the ecosystems around them in space, materials, and time. Organic architecture needs to work spatially with its surroundings, practically as well as aesthetically; it needs to use materials structurally as well as aesthetically suited to their purposes, and produce, employ and dispose of those materials in ways that do no harm to natural systems; and it needs to adapt over time to the needs of the people who use it and to changes in the natural systems that surround it. Once the age of cheap abundant energy is over, an architecture that meets these requirements will offer more efficient use of limited resources than the architecture of extravagance that defaces so many landscapes today.

Getting to an organic architecture of the sort suggested here may sound like a tall order. Still, like the other requirements of the ecotechnic future, it does not need to be done all at once, and important steps toward these goals are already being made. Just as the alternative agriculture movement of the 20th century yielded a range of techniques for organic food production that are already moving into niches vacated by industrial agriculture, similar alternative circles have pushed forward the evolution of a variety of sustainable housing technologies.

Three themes — materials, energy and ecology — define the directions along which this evolution has proceeded and large overlaps between these themes provide a broad unity to the movement. Cob building is among the most popular examples just now. Cob is a mixture of clay and straw, not too different from the blend still used to make mud bricks in many parts of the nonindustrial world today. Thus it turns extremely cheap and readily available raw materials into sturdy walls that make excellent thermal mass for solar heating arrangements. Straw bale building is another popular technology in the same circles, and in fact straw bale walls are usually faced inside and out with cob, to produce a stable, weatherproof wall with high insulation values — on average, about three times the heat resistance of a standard stud-built house wall.[4]

A second line of advance focuses on designing houses that meet at least some of their own energy needs. Passive solar heating, in which sunlight is absorbed by a variety of methods and used for space heating, and solar hot water heating, in which collectors of various kinds soak up sunlight and use it to provide domestic hot water, are two of the best developed technologies along these lines, and only the artificially cheap energy costs of the last quarter century have kept them from widespread use. Masonry stoves and other efficient ways of using firewood have also seen much development, though these need to be combined with coppicing or other methods of sustainable woodlot management to yield an ecologically functional system.[5]

Finally, a handful of innovative thinkers over the last thirty years or so have embraced the idea of home as human habitat on the broadest possible scale, and designed homes that provide many of the basic needs of their inhabitants: growing food, producing energy, and taking care of waste. Classic examples are the Arks designed and built by the New Alchemy Institute, a group active in the 1970s and 1980s, and the Earthships of modern vernacular architect Mike Reynolds.[6] These combine excellent insulation, passive solar space heating and solar hot water generation with food-producing solar greenhouses and aquaculture tanks, composting toilets and other built-in conservation measures. The result is a home that functions as an ecosystem, meeting many of the needs of its inhabitants with minimal impact on the surrounding environment.

Projects such as these offer a great deal of hope that homes can evolve in an ecotechnic direction, and extend Wright's concept of an organic architecture in directions he himself never envisioned — which is, after all, the natural destiny of great creative insights. In the near term, though, other concerns will likely play a much larger role in shaping the homes in which most of us will live. Decisions made in the recent past have saddled most of the industrial world with a very different kind of architecture and used available resources so profligately that our choices will be sharply limited for quite some time to come.

retrofitting the future

Social critic James Howard Kunstler has described the manufacture of suburbia as the single greatest misallocation of resources in the history of the planet. The evidence suggests that he is not exaggerating.[7] Starting in the 1920s, and shifting into overdrive in the 1950s, the whole shape of America's human geography was reworked to fit the needs of automobiles, and in the process much of it lost any suitability for people. The same transformations spread a little later, and on a less gargantuan scale, to most other industrial nations. The result is a human landscape utterly dependent on the availability of energy cheap and concentrated enough to make a car-centric lifestyle available to most people. Since this condition is coming to an end around us, the suburban landscape as it exists today is a geography without a future.

This hard reality has inspired its share of fantasies about tearing the suburbs down and rebuilding some saner human ecology in their place. Unfortunately, the limits imposed by the predicament we're in put that pleasant hope out of reach. The resources that might have accomplished the task went instead into the quarter century of excess that finished off the age of cheap oil. Other demands for resources in the decades ahead of us will inevitably take precedence over the massive re-engineering project needed to replace the suburbs, or urban and rural landscapes that duplicate its car-dependent features. It may be possible here and there for new housing along more ecologically sound lines to be built, but for the time being, with these limited exceptions, the housing stock we have is the housing stock we are stuck with.

Limited as they are, the exceptions are crucial for the longer term. Today's housing stock will not last forever. Though much of it may have to endure for a good deal longer than its makers intended, very little will be around a hundred years from now; few of today's buildings have the durability that made the temples of Tikal viable long-term housing for survivors of the Classic Maya collapse. As subdivisions and condominiums become salvage sites, they will have to be replaced, and the ability to replace them with sturdy, comfortable and sustainable housing will be a top priority. The skills and knowledge base for ecologically sound building are widespread in alternative circles just now and with any luck these will become the foundations of a growth industry, as organic agriculture has, once we no longer have the cheap abundant energy that makes today's unsustainable architecture look viable.

Still, shortages of resources and energy in the immediate future make it improbable that more than a very small fraction of North Americans will dwell in a sustainable human habitat any time soon. In the wake of the recent housing bubble, many of the world's industrial nations already have many more houses than they need, while the economic system that supports home construction is so badly broken that it may take a generation to get a new one up and running. Thus most people will have to get by with existing homes and make them work in a energy-poor world by retrofitting.

The word "retrofit" was coined in the 1950s, but its place in common parlance is a legacy of the energy crises of the 1970s. During those years, a great many homeowners discovered that houses built to rely on cheap energy lost most of their advantages when energy stopped being cheap, while soaring interest rates and stagflation made buying a new home less viable than it had been during the preceding years. Many people responded by finding cheap, effective ways to improve the energy efficiency of existing homes. Insulating blankets found their way around hot water heaters, caulk guns traced lines around leaky foundation plates, insulated Roman blinds replaced fashionable curtains and a surprising number of people discovered that it really is just as comfortable to put on a cardigan as it is to turn up the thermostat on a cold evening. These changes and others like them can be made in today's homes as energy prices rise and the cost of waste stops being negligible.

These simple retrofits can be supplemented with more ambitious projects. Adding insulation is often the cheapest way to cut heating and cooling bills, usually the largest energy cost for a home. Solar hot water heating can be added to most homes with good results; even when a solar system cannot provide all of a home's hot water, it can preheat water going to a conventional heater and cut costs significantly. Flat plate solar collectors for space heating can be installed on homes that get winter sunlight, slashing heating bills, and solar greenhouses and sunspaces can be added to south-facing walls to provide heat and allow food production. Where wind blows steadily, wind turbines can supply electricity, and micro-hydro — small-scale hydroelectric systems that divert only a small fraction of the water from a creek or river — can be put to use where conditions allow. All these options, and more, have much to offer as today's economy of abundance gives way to the scarcity economics of the near future.[8]

the household economy

As mentioned earlier, though, the transformation of the home from an energy sink to an energy source is only part of the challenge we face in dealing with existing housing. The transformation of the home back into a productive economic unit imposes requirements of its own. These involve extensive changes to the physical structure of homes, making room for productive activities, but the changes in attitudes and behavior required may be more sweeping still. The best way to start exploring those is to talk briefly about an errand that is a routine part of my life as I write this, and will become part of many other lives as the economy of abundance runs down.

In the small southern Oregon town where I lived until recently, Tuesday is farmer's market day, and every Tuesday between March and November, my wife Sara and I used to walk three-fourths of a mile to the National Guard armory parking lot, where local growers and ranchers sell their produce. Among our regular summer purchases was a flat of fresh raspberries, which we turned into home-canned raspberry jam for the year to come.

It's true that we could buy the same volume of commercially manufactured raspberry jam and eat that instead. Still, these two ways of putting by a supply of raspberry jam are by no means equal. Set aside for a moment the higher quality of homemade jam, made of fresher ingredients and prepared in small batches; a central issue is that the homemade jam represents a much more efficient use of fossil fuels.

To begin with, the grower who produced the raspberries used organic methods, which saved the petroleum and natural gas that would otherwise have gone into pesticides and fertilizers. While she used a pickup to get to the market, the ten miles or so she drove compares favorably to the thousands of miles farm products are routinely shipped in their journey to factory, warehouse and supermarket. Turning berries into jam and canning the result probably takes about an equal amount of energy per pint of jam whether it's done in a home kitchen or a huge factory, though the energy can be provided more easily via a solar cooker or other renewable source on the home scale. Even without that, though, the homemade jam takes a small fraction of the energy to go from raspberry canes to pantry that commercial jam requires. One measure of these energy economies is that, including all expenses, our homemade jam costs us only about two-thirds as much as a commercial jam made of the same ingredients.[9]

From the viewpoint of conventional economic thinking, though, in terms of its impact on the gross domestic product — generally considered the broadest measure of national prosperity — our homemade jam is an economic disaster. Modest spending on raspberries, sugar, pectin and new lids for our much-recycled canning jars are the only contributions it makes to the economy. By contrast, making, shipping, storing and selling commercial jam requires, directly and indirectly, the expenditure of plenty of money, all of which counts toward a higher gross domestic product.

Consider the economics from the perspective of the people who make the homemade jam, though, and things take on a different shape. Aside from other reasons Sara and I might want homemade jam, we have a potent economic motive: by making jam ourselves we get a superior product at a lower price. The raspberry grower, in turn, benefits handsomely from the same decision; the price she gets for berries sold directly to the consumer is several times the price offered by wholesalers. According to conventional economics, the end result of individuals freely pursuing their own interest in a market should be the maximization of prosperity — and yet if prosperity is measured by the gross domestic product, our free pursuit of our own interest decreases our contribution to national prosperity.

This reflects one of the largest blind spots of contemporary economics: the assumption that market transactions mediated by money are the only significant economic activity. Our household jam-making activities drop off the economic radar screen the moment we finish paying for the raw materials. Value is being produced — the same jam offered for sale at next week's market would bring substantially more than the cost of the raw materials — but it's produced outside the market economy and therefore has no official existence in an economy measured entirely by market metrics.

What makes this particularly relevant in the twilight of the age of cheap oil is that the world's industrial nations, and above all the United States, have spent most of the last century moving as much as possible of the household economy into the market sphere. In making our own jam, Sara and I belong to a minority of American households. Glance back a hundred years, by contrast, and nearly every family outside the very rich and the very poor had an active household economy that produced a large fraction of the goods and services they consumed. Many factors contributed to this dramatic shift, but the most significant, once again, is the availability of cheap abundant energy.

Most of the economies of scale that make mass production of processed foods economically viable, after all, are economies only because of the low cost of transportation. As recently as the first half of the 20th century, most consumer products in the US were produced locally for regional markets because transportation costs were still high enough to make national distribution a costly proposition. (Those brands that did find a nationwide niche, such as Coca-Cola, did it by franchising their manufacturing and packaging to local firms.) It took a new transportation network of diesel-powered trucks using a massive interstate highway network to create today's national distribution chains, and cheap petroleum provided the foundation on which the whole system rose.

The twilight of cheap oil, in turn, bids fair to throw this economic centralization into reverse. As transportation costs become a major part of the cost of consumer products, the economies gained by local production will sooner or later outweigh economies of scale, opening economic niches for small firms nimble enough to move with the currents of change. Equally, though, the advantages of the household economy will become overwhelming. In a world of expensive energy, anything that can decrease the need for fossil fuels will pay off handsomely and with the coming of scarcity industrialism the choice faced by households throughout the industrial world may well come down to doing things themselves or doing without.

the decline and fall of home economics

Raspberry jam is one of hundreds of goods and services that until recently were produced in the household economy, outside the reach of the market. Nowadays nearly all those goods and services are either produced commercially or are not available at all. That represents an economic and social transformation of massive scope.

That transformation can be measured by a rarely noticed fact. Visit the library of any American university that has not yet taken up the fashionable habit of purging its collection of "outdated" materials, wander through the stacks until you find the most neglected shelves in the building, and odds are that you'll be gazing on the mummified remains of a field of study, a profession and a university department as dead as the dinosaurs, and a good deal less popular nowadays: home economics.

Not that many decades ago, an impressive network of home economists working for universities, county extension services and private industry provided an extensive support system for the household economy. Backing that network was a consensus that recognized the social and economic importance of the household economy. That consensus was supported by the experience of two world wars, in which government-promoted home economics measures played a major role in softening the impact of food rationing and enabling the United States to feed armies and allies alike.

At the same time, the household economy faced steady pressure from the expansionist drive of the market economy. Beginning around the end of the 19th century, the market seeped into the domestic sphere with a steady stream of "convenience" products and "labor-saving" devices. The emergence of home economics as a profession and a scholarly discipline was itself partly driven by market forces, as businesses and governments looked for ways to enlist the household sphere in the market economy. These two social forces — the consensus supporting the household economy and the expansion of a metastatic market economy — collided head on in the decades after the Second World War. A third force, however, played the decisive role in the collision. That force was the role of the economics of gender in shaping the second wave of American feminism in the 1960s and 1970s.

Many currents of social change flowed together to launch the women's movement of the 1960s, but one factor not always given its due is the abrupt changeover from the war economy of the 1940s to the consumer economy that followed. As the troops came home, government and industry alike did everything in their considerable power to get Rosie the Riveter off the factory floor and turn her into Suzy Homemaker as fast as possible to free up jobs for demobilized soldiers. At the same time, the quest for markets to fuel the consumer economy's expansion and employ those same millions threw the market assault on the household economy into overdrive.

Postwar propaganda—"advertising" is too mild a word for the saturation campaigns that flooded popular media in the late 1940s and early 1950s—deluged middle class families with glittering images of affluence in which convenient, up-to-date consumer products provided by the market would replace the dowdy routine of the domestic economy with a life of elegance and leisure. The reality behind the image turned out to be much less palatable. Denied both the place in the market economy they had occupied during the war years, and the role in the household economy their mothers had held before that, millions of women across America found themselves expected to lead a purely decorative and purposeless existence.

As a motor for rebellion, deprivation of meaning is even more potent than deprivation of food, and so an explosion was inevitable. Many of the forms that explosion took were altogether admirable. Many injustices were set to rights, or at least challenged, and social roles that had become hopelessly restrictive for women and men alike came in for a much needed reassessment. Still, as the feminism of the Sixties and Seventies percolated outward into popular culture, it suffered the common fate of progressive social movements: instead of challenging the system of male privilege and the presuppositions that underlay it, many women who considered themselves feminists simply set out to seize their share of privilege within the existing order.

In that process, many of them embraced the manners, mores and attitudes of the privileged group they hoped to enter. Compare a 1960s *Playboy* with a *Cosmopolitan* from the 1980s or 1990s, for example, and it's impossible to miss the parallels, from the shared obsession with sexual conquest, conspicuous consumption and personal appearance, to the interchangeable cover girls meant to allure potential readers. The astonishing thing is that the "Playboy man" and the "Cosmo girl," those airbrushed icons of mindless consumer culture, were both considered to be liberated and liberating in their day.

The household economy, or what was left of it, was one of the casualties of the process that made these dubious figures popular. The feminist movement might have posed hard questions about the relative value assigned to household and market economies, and indeed some of the deeper minds within the movement made forays in this direction, but their ideas found few listeners. Instead, many feminists — and, eventually, many American women — simply accepted the relative values their culture assigned to the two economies, and aspired to the one they were taught to consider more valuable. That shift in attitudes cut the ground out from under what remained of the consensus that gave home economics its place; by 1980 most universities had closed their home economics departments and county extension agencies and private firms followed suit.

Still, the economic roles once assigned to women in the household economy had become so heavily burdened with social meaning that the images connected with them had to go somewhere. To a remarkable extent, they were applied to the institution that supplanted the economic roles once held by women: the market itself. Look at the rhetoric applied to the market in recent decades and you'll find every cliché applied to women present and accounted for. The market is America's seductive sex kitten offering tantalizing pleasures, its June Cleaver mom decked out with patriotic flags and apple pie and its nubile innocent waiting to be rescued from the lustful grasp of government bureaucrats and tax collectors. Placed on a rhetorical pedestal as florid as anything Coventry Patmore ever said about Victorian womanhood, and exploited as ruthlessly as Victorian women so often were, the market is America's pinup girl, the focus of overheated notions every bit as detached from real life as the fantasies that filled *Playboy* or *Cosmo* in their prime.

Any attempt to rebuild the household economy in the wake of peak oil will have to contend with these issues. It's not uncommon today, for example, to find couples for whom professional childcare, an extra car and commuting expenses, and the other expenses of a two-salary lifestyle cost more than the second salary brings in. Many of these families would come out substantially ahead if one adult were to stay home and work in the household economy, providing childcare and the like at a sharply lower cost to the family budget, but in the present social climate, this option is unthinkable for many people.

As a longtime househusband, I can speak about this from experience. For a good half of my married life, it made more economic sense for Sara, a bookkeeper, to work in the market economy, while I tended the garden, cooked the meals, did most of the cleaning and worked my way through the learning curve of a writing career in my off hours. I came in for a fair amount of criticism for making this choice, though it was a great deal less savage than the treatment meted out, mostly by other women, to women I knew who made the same choice. Despite the social pressure, though, it was the right choice for us, and enabled us to maintain a comfortable lifestyle on a modest income.

That choice is likely to be a valuable option for many more people as the market economy contracts in the wake of peak oil. The abandonment of the household economy, after all, happened because of the temporary conjunction of American imperial expansion with the massive fossil fuel production of the postwar years. As America's empire frays and global energy supplies falter, the costs of the energy-intensive economic structure built over the last sixty years will rapidly outweigh its benefits. In that context a renewal of the household economy offers one valuable set of tools for taking up the slack and providing needed goods and services. Those dusty books in the home economics section of the local college library may become valuable once again.

Such a renewal, though, will require a reassessment of social roles and values as ambitious as anything the feminists of the 1960s envisaged. Measures of value evolved within the market and shaped to a large degree by market-centered ideologies fall flat when applied to nonmarket economies in which custom, reciprocity and collective benefit govern exchanges. Money itself, that abstract fiction that has very nearly smothered the real economy of goods and services it evolved to support, may be less relevant as alternative forms of value become ascendant. The forms taken by those alternatives in the ecotechnic world of the future is probably impossible to guess at this point, but an openness to options and a willingness to look beyond the market are likely to be valuable steps just now — and a renewed household economy is a seedbed in which the economics of the future can take root and grow.

the specialization trap

The reinvention of the household economy will also be a stark necessity as the age of abundance capitalism gives way to later seres in the successional process of the future. The centralized production and distribution of goods that defines the industrial economy is vulnerable to breakdown, and not only because it uses extravagant amounts of fossil fuel energy. An example from the past shows where the dangers lie.[10]

In the Roman Empire, ceramics were a central technology and the Roman pottery industry was huge, capable and highly centralized, churning out tableware, storage vessels, roof tiles and other goods in such quantities that archeologists across Roman Europe struggle to cope with the fragments today. The pottery works at La Graufesenque in southern Gaul, for instance, shipped exquisite products throughout the western empire and beyond it. Ceramics bearing the La Graufesenque stamp have been found in Denmark and eastern Germany, hundreds of miles past the Roman frontiers. Good pottery was so cheap and widely available that even rural farm families could afford elegant tableware, sturdy cooking pots and watertight roof tiles.

All this ended when Rome fell. When archeologists opened the grave of a sixth-century Saxon king at Sutton Hoo in eastern Britain, the pottery they found told a stark tale of technological collapse. Had it been made in fourth century Britain, the Sutton Hoo pottery would have been unusually crude for a peasant farmhouse; two centuries later, it sat on the table of a king. What's more, most of it had to be imported because the potter's wheel dropped entirely out of use in Britain – one of many technologies lost in a cascading collapse that took the island down to levels of impoverishment more extreme than anything since the subsistence crises of the middle Bronze Age more than a thousand years before. Cooking vessels, food containers and roofing that keeps out the rain are basic to any form of settled life. An agrarian society that cannot produce them is impoverished by any definition; an agrarian society that had the ability to produce them, and then loses that ability, has undergone an appalling decline.

What makes this example relevant to the present is that the post-Roman collapse had its roots in the sophistication and specialization that made the Roman economy so efficient. Huge pottery factories like the one at La Graufesenque, which used specialist labor to turn out quality goods in volume, could make a profit only by marketing their wares across much of a continent, using far-flung networks of transport and exchange to get products to consumers who wanted pottery and had denarii to spend. The Roman world was rich and stable enough to support such networks — but the post-Roman world was not.

The political implosion of the Roman Empire thus turned an economic advantage into a fatal vulnerability. As transport and exchange networks came apart, the Roman economy went down with it, and that economy had relied on centralized production and specialized labor for so long that no one knew how to replace it with local resources. During the Roman Empire's heyday, people in the towns and villas near Sutton Hoo could buy pottery from local merchants who shipped them in by land from southern Britain and by sea from elsewhere in the Empire. They had no need of local pottery factories, and so nobody there knew how to make good pottery. This meant that their descendants very nearly ended up with no pottery at all.

Even where Roman pottery factories still existed, they were geared toward mass production of specialized types of pottery, not small-scale manufacture of the whole range needed by communities. Worse, as population levels plunged during the Roman collapse, the salvage economy that always springs up in the wake of collapse made use of the abundant stock of pottery on hand, removing any market for new production. A few generations of economic contraction, social chaos and depopulation thus erased the craft traditions of Roman pottery-making, leaving the descendants of potters with no knowledge of how to make good pottery.

Trace any other economic specialty through the trajectory of the post-Roman world and the same pattern appears. Economic specialization and centralized production, the core of Roman economic success, left Rome's successor states with few choices and fewer resources in a world where local needs had to be met by local production. Caught in a trap defined by their own specialization, most provinces of the former empire entered the Dark Ages more impoverished than they had been before the Roman economy evolved in the first place.

The same process is likely to shape the twilight of the industrial age. In modern industrial nations, the production and distribution of goods have become far more centralized than anything Rome ever achieved. Nearly all workers in the industrial economy perform highly specialized niche jobs, most of which depend on an equally centralized, mechanized and energy-intensive global economy, and many of which have no meaning or value outside that economy. If the structure comes apart, access to even basic goods and services could become a challenge very quickly.

Food is the obvious example. Only a small number of people in any industrial nation know how to grow their own food and not all of those have access to the land, tools and seed stock to give it a try. The same principle holds, however, for every other necessity of life and countless other things that would be good to have in the deindustrial dark ages. Consider the skills needed to find and process useful fibers, spin and weave them into cloth and make the cloth into clothing. Few people have any of those skills, much less the entire set; the tools needed to do most of them are hardly household items these days and building and repairing those tools takes specialized skills. Our situation is thus far more precarious than Rome's.

This is where the household economy again comes into its own. Escaping from the jaws of the specialization trap requires that individuals learn, practice and pass on the skills that have been replaced by mass production. The household economy is the best place to get this necessary process under way. The craftspersons of a revived household economy will not have to produce all the goods and services they need at first. Rather, as scarcity industrialism replaces the economy of abundance, they will fill much the same niche as the small market gardens in the former Soviet Union: at first, hobbies; next, sources of additional income in a time of declining prosperity; finally, a desperately needed lifeline providing most of the goods actually used by people as the larger economy comes unhinged.

These concerns, finally, reach back to the issues of home design raised earlier in this chapter. It is not enough to build the homes of the future or retrofit those of the present so they use as little fossil fuel as possible. Those houses must also provide space for the household economy. Compare the roomy and practical kitchens of old farmhouses with their poorly designed and cramped equivalents in most recent houses, and some of the issues involved will be clear. The need to make homes function as production zones imposes other requirements as well. Good lighting is essential for many crafts and this has to be considered when planning an energy-efficient home. Equally, the household economy includes as many things around the house as inside it — for example, a garden for herbs and vegetables, including a solar greenhouse or cold frames, outdoor tool storage and work space for crafts best done in the open air or in an outbuilding. All this will shape what can be done with the space surrounding the home.

All these factors need to be kept in mind by families and communities as they face up to the challenge of getting ready for the transition into the deindustrial age. On a family level, retrofitting or new construction can respond to the need to reinvent the household economy, and those who take the time to learn the skills of a household economy will likely find that the investment pays rich dividends in the years to come. On a community level, modest changes in zoning regulations and city ordinances — for example, rules encouraging home businesses, farmer's and craft markets and the like — could contribute mightily to the birth of a domestic economy better suited to the approaching deindustrial age. One way or another, the household economy will be needed, because the changes the future will impose on the economy of paid employment and market exchanges may well leave few of today's economic patterns intact.

Work 8

Much of the work that goes into keeping individuals, families, and communities alive will thus have to be done in the household economy. Still, work outside the home will continue to play an important economic role. The form of market economy that controls so many flows of goods and services just now is unlikely to survive long and market economies in any form will likely become a good deal less significant than they are today. Still, markets have been around for a long time and are not likely to go away soon.

The age of scarcity industrialism, in particular, promises to birth a distinctive form of market economy, different from the one that provides most people with their livings today but not so different from the one that employed most of our great-grandparents. It has only been for the last few generations that energy has been so cheap that building and powering a machine makes better economic sense than training and employing a human being. As fossil fuels deplete, this equation will often reverse.

The implications of this change render most current discussions about economic development meaningless. Until recently, talk about disparities between the world's industrial societies and the rest of the planet focused on how to bring the Third World "into the twenty-first century." The phrase itself betrays the ideology that shaped that discussion — the beliefs, as devoutly held as any other religion, that history progresses straight to us, that any other social arrangement is simply some version of our own outmoded past and that our peculiar way of managing human communities is thus as inevitable as it is necessarily beneficent.

This whole discussion was an exercise in futility from the start. The economic crash that arrived in 2008 showed all too clearly that when people in the world's non-industrial societies did exactly what so much rhetoric insisted they ought to, and claimed a share of the world's fossil fuels and industrial output for their own, the result was soaring energy costs and economic chaos. The limits to growth were always there; it was merely the political arrangements that restricted the benefits of industrialism to a small portion of our species that made it look as though unlimited growth was ever an option.

A central fact at this juncture of history is that the way things have been in the world's industrial societies over the last century or so is wildly abnormal by the standards of the rest of human history. The transition away from that abnormal experience to a more normal one — from an economy of abundance to one that is limited by scarce and expensive energy — is beginning now. That transition will change nearly everything in our lives, but few things will be affected as powerfully as the realm of work and employment. Too little attention has been paid to the question of how people will earn a living during the descent into the deindustrial future. In the centuries to come the world will change dramatically, but there is no reason to think that these changes will make earning a living easier than it is today.

a hundred energy slaves

The huge distortions that industrial nations underwent due to the flood of cheap energy that washed over them in the 20th century can be measured by a simple statistic. In the United States today, the amount of energy used each year works out to around 1,000 megajoules per capita, or the rough equivalent of 100 human laborers working 24-hour days for each man, woman and child in the country.[1] The bill for all this energy comes to around $500 billion annually, or about $1,667 per person per year.

This may seem normal, until one asks how much it would cost to hire human laborers to perform the same amount of work. At the current US federal minimum wage of $5.75 an hour, hiring three shifts of 100 workers each to provide the equivalent amount of energy would cost each American $512,811 a year, or some 308 times the energy cost — and this ignores payroll taxes, health insurance, paid vacations and the like. Mind you, it would also require the US to find food, housing and basic services for an additional workforce of 90 billion people, but we can abandon the analogy before tackling issues on that scale.

The labor of these virtual guest workers has had a huge impact on lifestyles throughout the world's industrial societies. As recently as a hundred years ago, at the dawn of the petroleum age, most work in the industrial West was still done by human beings using hand tools. Kitchens had servants instead of appliances, factories and shops had workbenches and hand tools instead of robots, and the functions now carried out by computers were performed instead by armies of clerks wielding pen and ink. Go back two more centuries, to the eve of the industrial revolution, and the equation that identified work with human labor was even more exact. Horses and oxen were used to pull wagons and draw plows, and wind carried ships across the sea and shared with water power such limited tasks as grinding grain and fulling cloth, but outside of these and a few other very limited applications, economic activity involved human muscles and minds.

The familiarity of our current arrangements, and the rhetoric of progress we use to justify those arrangements, make it easy to dismiss such a human-powered economy as a primitive oddity that existed only because people didn't know any better. Look at the disparity in economic terms, though, and a different picture emerges. In any society not awash in cheap energy, it usually makes more economic sense to train and employ a human worker than to develop, build and power a machine to fill the same job. The machine may be cheaper in the abstract, even in the face of rising energy costs, but a machine does not exist by itself. It gets its economies of scale because it participates in a complex system that manufactures, distributes and powers it, and gives its work economic meaning. Lacking that system, many machines quickly lose value — how useful is a computer without cheap power, readily available maintenance and spare parts, and a society set up to give value to the narrow range of productive activities a computer can do?

This was why ancient Rome and imperial China, both of which had a solid understanding of mechanical principles and sophisticated technical traditions, never had industrial revolutions of their own. Lacking any way to access the fossil fuels that made modern industrial civilization possible, any attempt by these societies to replace human labor with machines would have been a failure from the start, and both sensibly put the available resources into expanding the labor force instead. The Romans did this the cheap, crude and ultimately self-defeating way, by pushing the growth of a slave economy to the breaking point. The Chinese did it far more sustainably and effectively by evolving a robust system of small-scale capitalism, on the one hand, and durable traditions of specialized craftsmanship on the other.

All this has a pressing relevance to our present situation, because the energy resources that make it possible for every man, woman and child in America to dispose of the equivalent of $512,811 in labor every year are running out. It's as though the 90 billion invisible guest workers whose sweat powers the American economy are quitting their jobs one by one and moving back to their homes in the Paleozoic. When the long curve of depletion finally sinks low enough that it's no longer economical to extract the last dregs of fossil fuel from the ground, the labor each of us will have at our disposal will be much less than it is today.

With any luck, it may be more than 1/308th as much — we know more about collecting energy from renewable sources than the Romans or the ancient Chinese did. Still, it is mere wishful thinking to assume that the trajectory of industrial expansion must continue into the future. There are plenty of claims nowadays that the universe is obliged to give us another vast windfall of cheap abundant energy to replace the one we've wasted so enthusiastically over the last few centuries but these are statements of faith, not of fact. Far more likely is the reemergence of an economy in which the work of human hands and minds is once again the main source of value.

With luck and hard work, the economy of the future may be a good deal closer to the Chinese than the Roman model. The farmers of the future may well use intensive organic methods rather than the field agriculture of an earlier day, just as some of the craftspeople of the future may well spend their time crafting solar hot water heaters and shortwave radios with hand tools. A handicraft economy is a mature and effective social technology and it can easily be adapted to the needs of the deindustrial future. This implies, among other things, that those who get the training needed to take part in a handicraft economy may be much better prepared for the future than those who pursue forms of employment that seem more attractive by today's standards.

The economic transformations that lead to such arrangements, like the rest of the deindustrial transition, will take place a step at a time. During the age of scarcity industrialism, something not too different from today's money economy will likely remain in place, though the household economy will play a growing role. During the age of salvage economies, in turn, money of some sort will likely remain in use on a small scale, but most day-to-day transactions will take place via barter or other systems of exchange outside the money economy. The twilight of money is standard in dark ages after the fall of a civilization. In both periods, though, people will work for a living — and they will likely work much harder than people in North America do today.

Their jobs will not be the same as those that employ most people in North America nowadays. Few people in the industrial world spend their work hours producing goods or providing services that people use. Instead, jobs focus on information management and the needs of the industrial economy itself and contemporary education focuses on preparation for a career pushing papers, typing data into computers or managing human beings. The once-mighty archipelago of trade schools that undergirded the rise of the United States as an industrial power accordingly sank from sight in the second half of the 20th century. I once lived a few blocks away from the shell of one such school, the Thomas Edison Technical School in Seattle. It had been engulfed by a community college, and classrooms once full of the hum of machine-shop equipment and the hiss of hot solder were being used to train receptionists, brokers and medical billing clerks.

Still, the post-industrial economy proclaimed by Daniel Bell many years ago, and accepted as economic reality by most people since then, was never more than a shell game.[2] The societies of the "post-industrial world" are just as dependent on industry as they have ever been; they simply export the industries to Third World countries where labor is cheap and environmental regulation nonexistent, and continue to reap the benefits back home. Those arrangements can only work for as long as cheap abundant energy makes transport costs negligible. It is only cheap fossil fuel, backed by the systematic distortions in patterns of exchange that pump wealth from the Third World to a handful of industrial nations, that provide the latter with the wherewithal to pay a large fraction of their populations to do jobs that don't actually need to be done.

the deindustrial want ads

As energy becomes scarce and expensive again, and the imperial systems that concentrated the world's wealth in a minority of nations are shredded by economic breakdown and the rise of new centers of power, those arrangements will come apart. As that happens, a great many necessary goods and services now produced offshore will need to be done at home once again, and a great many professions that provide no necessary goods and services will likely drop off the economic map for a long time or forever. The result will reshape the economies of most industrial nations, and once again those who are ready for the resulting changes are much more likely to make the transition more successfully than those who are not.

Prophecy is a risky business, but it may be worth hazarding some guesses about the jobs that will fill the want ads here in North America over the next generation or so, as scarcity industrialism takes shape. Farmers are among the most likely candidates for the top of the list. By this I don't mean subsistence farmers in rural ecovillages — their time is much further in the future, if it ever arrives. Rather, market farmers capable of tilling what is now suburban acreage to feed the dwindling cities and rural farmers who can produce grains and other bulk crops for foreign exchange will likely be in high demand, along with support professions such as agronomists and organic farming instructors.

Many of the professions that play roles in the baroque food chain that separates farmers from consumers, on the other hand, are likely to be early casualties of the transformations ahead. Only huge energy inputs make it possible to support today's network of factories, warehouses, trucking firms, commodities markets, advertising agencies, supermarkets and the like. In the age of scarcity industrialism, as dwindling fossil fuel reserves have to be prioritized for urgent needs, none of these can count on having access to the energy they need and most will go out of business. Their replacements, though, will rank high on the list of growth industries. Individuals and small firms that can turn locally available raw materials into locally valued products will do well. This includes foodstuffs, of course, but also such necessary products as soap and certain luxuries as well. Those who can brew good beer, for example, will be able to find a lively market for their skills even (in fact, especially) in the most difficult of times.

Engineering is another trade likely to do well during the age of scarcity industrialism, especially the fields focusing on energy production and distribution and low-tech transportation networks. In the not too distant future, rail and canal transport will have to take over much of the work now done by trucks and energy networks will have to cope with a fractious mix of alternative resources, dwindling fossil fuels and conservation programs. Those who learn enough about natural cycles and the realities of biology to craft sustainable systems in any field will likely find their services in high demand. Meanwhile, the craftspeople and laborers who put the plans of engineers into effect, from skilled machinists all the way down to the gandy dancers who lay the rails, can also count on steady paychecks.

One group of professions likely to prosper in the age of scarcity industrialism, and even more so in the salvage economies that will follow, barely exists today: the salvage trades. Demolitions experts, junkyard workers, small appliance repairpersons and people who run recycling and composting operations represent the first forays into the territory, but many new professions will emerge in this field in the years to come. A huge fraction of North America's potential wealth in the post-peak years will consist of manufactured objects from the recent past that can either be refurbished and put back into circulation or stripped for raw materials. When electricity can no longer be spared to power elevators and run heating and cooling systems, for example, skyscrapers will be worth more as sources of refined metal than as buildings, and they will start to be torn down. When the cost of repairing a broken toaster once again becomes cheaper than buying a new one, repair shops will become paying businesses again, as they were half a century ago. The salvage trades may thus turn out to be among the major growth industries of the 21st century.

Another group of professions that exist only in outline today, but will likely be major growth industries in the future, are those trades that work at the interface between human society and the natural world. The brute force approaches to environmental problems used today, from damming rivers for flood control to poisoning unwanted insects, are only possible because the energy and resources needed to enact them and deal with their unwanted consequences are still abundant. Future societies that need to resolve problems in a less extravagant manner will find it much more cost-effective to use subtler ecological interventions: solving flood problems with plantings and swales that absorb runoff, for example, or curing insect infestations by making the local habitat less inviting to pests and more attractive to predators. The trades that work along these lines — call them the ecotechnic trades — will draw on perspectives now scattered among ecology, agronomy, geology, public health and a baker's dozen of other specialties in science and engineering. Their emergence as recognized professions may take longer than the salvage industries, but the need for them already exists.

The twilight of the information revolution offers other possibilities. Long before economic decline shreds the industrial infrastructure needed to build and maintain today's computers, their direct and indirect energy costs will restrict computers to a narrowing circle of government and business uses. As energy costs outstrip labor costs, secretaries and bookkeepers may once again be more economical than word processors and spreadsheets. In past societies, most calculation, analysis and memory work was done by trained people, whose abilities in these fields were as far beyond today's usual standard as professional athletes today surpass the capabilities of the average urban couch potato.[3] It may be some time before these skills emerge as the focus of established professions, but steps in that direction may not be out of place right now.

Other professions have their own possibilities. It's a useful exercise to locate a city directory from the first half of the 20th century and flip through the pages, noting the businesses that no longer exist. Those that still meet the needs of individuals and families — the butcher, baker and candlestick maker of the old nursery rhyme come to mind — however unpopular they are as career tracks today, will likely be more viable and more lucrative in a deindustrializing future than many currently fashionable professions. Pundits never seem to tire of explaining that tomorrow's jobs will not be the same as today's; they may be right, but many of tomorrow's hottest jobs may resemble the profitable careers of a century ago. Those people who make preparations now to move into such jobs as they come open will be doing themselves and their communities alike a favor.

Some attention should also be given to skills that will be useful further down the curve of decline and fall. To some extent this is a matter of preserving knowledge that would otherwise have to be reinvented decades or centuries from now, but in some cases it may also have immediate value. The end of the age of abundance industrialism means the end of the trickle-down economy that, despite its massive faults, provided a great many economic benefits to the middle classes and raised the industrial world's working classes out of abject poverty. To some extent, while the privileged classes will be entering a new order of scarcity industrialism, people outside those classes may find themselves passing directly into the world of Dark Age salvage societies. What this implies is that for many people the skills and habits of the age of salvage may be well worth cultivating now.

For others, skills suited to the age of scarcity industrialism will likely pay for themselves immediately. One telling phenomenon of the Seventies was the emergence of home energy retrofitting as a viable profession. In every American city and a great many smaller towns, contractors no longer able to find work building houses found a new niche installing insulation, storm windows and solar water heaters, while hardware stores found space for a new section of home energy efficiency supplies. It was never a large sector, and its growth came to a sudden stop in the early 1980s in the flurry of political machinations that crashed the price of oil and threw away our best chance for a transition to sustainability, but it was one of the few success stories at a time when most American industries were contracting and most families saw their standard of living slipping year after year.

Many of those same conditions are repeating themselves on a much larger scale as the world moves into the territory beyond peak oil. While recent swings in the price of energy make it hard to see the larger picture, the general trend in energy costs is still up; one step down for every two steps up still amounts to steady upward movement. The economic crisis now beginning may well make the stagflation of the Seventies look mild, but to many families it poses the same challenge of having to get by with less.

The auto industry is facing a similar transition as mechanics and hobbyists across the country turn used cooking oil into biodiesel, convert hybrid cars into plug-in vehicles and equip bicycles and scooters with electric motors and batteries. If the Detroit auto companies survive long enough to get their much-ballyhooed electric cars to the market, they may find themselves eating the dust of thousands of ingenious retrofitters who, lacking the institutional inertia of Fortune 500 corporations, are getting products to local markets right now. These retrofits won't allow what James Howard Kunstler has usefully labeled "the paradise of happy motoring" to continue. On the other hand, they may well enable many Americans to deal with the downside of a social geography designed for cars rather than people during the inevitable lag time before that social geography becomes a bad memory.

Many more dimensions of American life will need retrofitting in the years to come. Nearly every aspect of our economy, culture and politics depends on cheap abundant energy and will have to be rebuilt to deal with the new reality of energy scarcity. That will apply to little things — for example, plenty of home appliances now controlled by computer chips can be made to work with thermostats, spring-driven timers and the like, given a little ingenuity and a willingness to tinker — as well as to massive ones. In a very real sense, given the sharp limits we face in the near future, our entire lives will need to be retrofitted to deal with the new realities of the deindustrial age. The first job of the salvage economy will be to haul a viable future out of the scrap heap of the present and get it back into some semblance of working order.

the twilight of automation

The same reversion to older patterns is likely to affect many more aspects of life in the near future, as changes that were proclaimed as the wave of the future turn out to be short-lived adjustments to the huge but temporary flood of cheap abundant energy that has shaped so much of our society. Technology offers a trenchant example.

Few aspects of contemporary life are as heavily freighted with mythic significance as the way that technologies change over time. It's from this, more than anything else, that the modern faith in progress draws its force. Yet there are at least two radically different processes lumped under the label of "technological progress." One is the actual development of new technology; the second is the replacement of human labor with energy slaves. The first of these begins with an unexpected leap — the birth of some new technology is rarely planned — but thereafter it follows a predictable course driven by the evolution of the underlying technology: the first clumsy, tentative and unreliable prototypes are replaced by ever more efficient and reliable models, until something like a standard model emerges. The resemblance between this pattern and the course of succession outlined in Chapter Two is intriguing and may not be accidental.

Once the technology reaches its "climax community" stage, variations are driven mostly by changes in fashion. Compare a sewing machine, a clothes dryer or a turboprop engine from the 1960s with one fresh off the assembly line today, for instance, and the differences in the underlying technologies are fairly slight. The great difference lies in control systems. The machines of the 1960s used relatively simple mechanical means of control, guided by the skill of human operators. Their equivalents today use complex digital electronics powered by energy slaves and require much less human skill to run effectively. On a 1960s sewing machine, for example, buttonholes are sewn using a simple mechanical part and the knowledge and skill of the seamstress. On many modern machines, the same process is done by tapping a few virtual buttons on a screen and letting the machine do it.

It's common to see changes like this lumped together with the process of maturation within a single technology. Yet there's a difference in kind. The change from mechanical controls and operator skills to digital electronics is not a matter of progress in a single technology. It marks the replacement of one technology for guiding a sewing machine by another. Mechanical controls and home economics classes did not gradually evolve into digital sewing machine controls; instead, one technology ousted another. Furthermore, both technologies do an equally good job of making a buttonhole. The factors driving the replacement of one by the other are external to the technologies themselves. In the case of the sewing machines, as in so many similar technological transformations of the last 60 years or so, these replacements furthered a single process — the replacement of human skill by energy slaves.

What drove this, in turn, was the same economic equation already discussed in this chapter: the fact that for a certain historical period, all through the industrial world, fossil fuel energy was cheaper than human labor. Anything that could be done with a machine was therefore more profitable to do with a machine, and the only limitation to the replacement of human labor by fossil fuel-derived energy was the sophistication of the equipment needed to replace the knowledge base and nervous system of a skilled laborer. For most people today, that equation still defines progress. A more advanced technology, by this definition, is one that requires less human skill and effort to operate. The curve of this kind of "progress" points to the fully automated fantasy future that used to fill so many comic books and Saturday morning cartoons.

One of the major mental challenges of the near future will consist of letting go of this fantasy and retooling our expectations to fit a different reality. It's hard to think of an aspect of modern life that will not face drastic reshaping in the course of this retooling. The collapse of American education, for example, was in large part a consequence of the same economic forces that put computers into sewing machines. For the last few decades, it was more cost-effective in the short term to hand over bookkeeping chores to computers and equip word processors with spell checkers than it was to teach American children how to do arithmetic and spell correctly. In the future, this will likely no longer be true, but the sprawling bureaucracies that run today's education industry are poorly equipped, and even more poorly motivated, to teach the skills that will be needed for humans to replace the machines.

Now of course not all the machines will need to be replaced at once. Many modern technologies, however, demand very large energy inputs that will not be reliably available in the future. Many more cannot be repaired when they break down — during the age of cheap energy it was more cost-effective to throw a machine away when it broke and buy a new model than it was to pay a repairman's wages, and so many machines in the last quarter century were not built to be repaired. Furthermore, the extraordinary levels of interconnection that pervade today's technology mean that the failure of a single component that cannot be replaced or repaired can render an entire system useless.

It's probably too late to avoid the future of systems failure the choices of the recent past have prepared for us, but quite a bit can be done to mitigate it. The first priority is to break free of the dubious assumption that the kind of technology that was more cost-effective in an age of cheap abundant energy will be well suited to the age of scarce and limited energy now beginning. The second is to redirect as much attention and effort as possible to technologies better suited to the new realities of our future.

trailing-edge technologies

Among the most useful resources in the first phase of the transition, in turn, are precisely the technologies that fell out of fashion in the last extravagant decades of the age of abundance. As a culture, we have pursued cutting-edge technologies for so long that shifting attention to trailing-edge technologies seems almost willfully perverse. Nonetheless, older technologies that work effectively with relatively modest energy inputs and rely on human hands and minds in place of energy- and resource-intensive electronics, may well turn out to be more viable in the long run than their allegedly more advanced replacements.

That 1960s sewing machine — designed to allow for maintenance and repair, assembled from easily replaceable parts and relatively easy to convert to foot pedal power if electricity becomes scarce — is likely to have a much longer working life in an age of decline than the computerized models filling showrooms today. In the same way, a great many trailing-edge technologies and the skills needed to use them, many of which can still be learned from living practitioners today, are worth preserving. The question, of course, is how many people will take on that project while the opportunity still exists.

Another example provides a glimpse at the scale of this opportunity. One of the great achievements of the last two hundred years, by most measures, is the communications network that sends information from one side of the planet to another at the speed of light. The telegraph, the first really successful long-distance communications technology, caused sweeping changes across the industrial world and launched a series of more complex media — telephone, radio, television and finally the Internet. Not all of these were an unmixed blessing. Every technology has its downside, but on the whole, widespread access to long-distance communication has been much more a blessing than the opposite.

At the same time, few dimensions of industrial society are more vulnerable to breakdown in the age of scarcity. The Internet, the crown jewel of modern communications, depends on a huge, energy-intensive infrastructure that will become impossible to keep running as energy and other resources become scarce. The Internet depends on many thousands of server farms, many of which use as much electricity as a small city, and the technology that makes the Internet possible in the first place requires abundant energy, exotic raw materials, costly and energy-intensive fabrication plants, hundreds of thousands of well-trained technicians and a social consensus that supports investing all these resources there and not elsewhere. Counting on all of these to remain available during the decades to come is a gamble against long odds. For that matter, most of today's communications media, including major segments of the

Internet, are paid for by advertising sales, and this model will be hard put to survive the end of the economy of abundance; economies that have to struggle to meet essential needs are unlikely to put scarce resources into encouraging people to buy more.

Yet electronic communications could be rebuilt along other lines, far less dependent on today's industrial system and its advanced technology. It's quite possible to build a vacuum tube — the backbone of communications technology in the days before transistors — from common materials using hand tools; Pete Friedrichs' excellent book *Instruments of Amplification*, which details how to do this, has become popular reading on the more exotic end of the do-it-yourself crowd in recent years. Fifty years ago, popular books for the teen market such as Alfred P. Morgan's *The Boy's First* (and so on up through *Sixth*) *Book of Radio and Electronics* taught aspiring young electricians how to build remarkably sophisticated gear out of oatmeal boxes, spare parts and salvaged scrap. Given the right knowledge base, it's not necessary to give up the opportunity to send messages around the world at the speed of light; it's just necessary to do it in a different and more sustainable way.

What might a viable communications net look like in the age of scarcity industrialism? To begin with, the structure of the network might best be a decentralized web of self-supporting and self-managing stations, sharing common technical standards and operating procedures. It would use the airwaves rather than land lines to minimize infrastructure and its energy needs would be modest enough that stations could get their power from renewable sources. It would use a broad mix of communications modes, so that operators could use the highest technology they had available but could scale down from there if necessary, all the way to low-power homebuilt gear sending Morse code. It would need a lively internal subculture to encourage technical skill and practical expertise in new members and to foster the sort of camaraderie that helps keep a far-flung technical network operating through difficult times.

This outline is hardly speculative; exactly such a network exists in the subculture of amateur radio.[4] Very few people outside the amateur community remember that, when radio transformed itself into a mass medium in the 1920s, the private experimenters who pioneered long range radio communications kept the right to use certain sectors of the airwaves and grew into a worldwide community of radio hobbyists who communicate with one another over a diverse assortment of technologies and bands. Licensed and occasionally encouraged by governments, "ham radio" — the origin of the nickname is a subject of some debate — flies almost completely under the radar of mainstream culture these days, surfacing only when someone in the media notices that in the wake of some natural disaster a group of local radio amateurs stepped up and kept emergency communications going when all other channels shut down.

As we move deeper into the age of peak oil, a network such as this has important practical advantages. First, as energy prices rise and technologies dependent on high energy inputs become less and less common, the Internet as it exists today is likely to be an early casualty. Versions refitted for government and military use may remain in place for some decades, but most ordinary users are likely to be forced offline in short order by rising prices and service cutbacks. Bands assigned to amateur radio, on the other hand, include several that allow worldwide contacts and others with regional or local ranges; the usefulness of such links in an age when other means of communication falter is hard to overstate. At the same time, the amateur radio community teaches the skills needed to build, repair and operate radio equipment. These skills have many applications outside the ham radio field; the potential for job training here is not small.

Second, many other technologies vulnerable to the impacts of peak oil and climate change have potential backups and replacements in the large and little-known world of hobby subcultures. An astonishing number of trailing-edge technologies, from black powder firearms through handloom weaving to long-distance sailing on windpowered boats, have become hobbies pursued by communities of aficionados. Historical reenactment societies are among the richest sources of these neglected technologies. Those communities and the knowledge they preserve could be an immense resource in the aftermath of the age of oil, as well as a significant source of job training for those willing to pursue it.

A third lesson, though, may be the most important of all. The first part of this book suggests that our civilization is the first, and the most clumsy and tentative, of a new class of human societies — technic societies — that are as distinct from urban-agrarian forms as the first urban societies were from the tribal cultures that preceded them. One of the inevitable blind spots our historical position imposes on us is a tendency to confuse the particular cultural forms evolved by our technic society with the requirements of technic societies in general. Amateur radio is a reminder that there are ways to handle long-distance electronic communications that do not involve massive high-tech infrastructure supported by huge energy inputs and the financial payback of a consumer economy. This has to be kept in mind as we begin the long transition toward the ecotechnic societies of a sustainable future.

Energy 9

Given that decreasing energy supplies define the most immediately challenging dimension of the crisis of industrial society, my readers may wonder why the subject of energy has been put off this long. Many other books on the consequences of peak oil focus primarily on energy issues. This one does not, because while energy is of crucial importance to today's industrial society, its impact on the individual is surprisingly indirect. Even in America, people do not eat, drink, wear or live in energy, and those who make a living producing and distributing energy are a distinct minority in the workforce.

Now of course it's true that the food we eat, the clothes we wear, the homes we live in and the jobs that pay our wages, among many other things, depend on a great deal of energy from fossil fuels. That dependence has become a massive vulnerability as fossil fuel energy depletes. To the extent that food, clothes, homes and jobs can be provided using less fossil fuel, or none at all, the potential for deprivation and human suffering can be lessened. Thus the critical steps needed to prepare for the future ahead of us, as discussed in Chapters Six, Seven and Eight, involve shifting from lifestyles that require plenty of cheap energy to ways of living free from that crippling dependence. Coming up with new sources of energy, in other words, is far less important than learning to use the energy we already have in a more efficient way.

When the collective dialogue of our time addresses energy issues, however, it fixates on the need to find some new source of energy to power the lifestyles we have now. This habit has deep historical roots. Several times already in the history of our civilization, technological developments — the steam engine, the electrical generator, the internal combustion engine and a few others — made it possible for human beings to use the huge reserves of fossil carbon stored inside the planet for their own purposes, and so opened the door that led to the industrial age. Thus it seems plausible to suggest that we can just repeat the process by coming up with more innovations that will give us access to more energy so that our current way of doing things can continue on into the future.

The only problem with this claim is that it is based on false logic. Imagine for a moment that instead of discussing the fate of industrial society, we are discussing an experiment involving microbes in a Petri dish. The culture medium in the dish contains 5 percent of a simple sugar the microbes can eat and 95 percent of a more complex sugar the microbes lack the right enzyme to metabolize. We put a drop of fluid containing microbes into the dish, close the lid and watch. Over the next few days, a colony of microbes spreads through the culture medium, feeding on the simple sugar.

Next, imagine that a mutation takes place: one microbe with variant biochemistry starts producing an enzyme that lets it feed on the complex sugar. Drawing on this more abundant food supply, the mutant microbe and its offspring spread rapidly, until finally the Petri dish is full of mutant microbes. Soon, though, the mutant microbes have eaten most of the supply of complex sugar and their survival is at risk. As we watch the microbes through a microscope, we might begin to wonder whether they can produce a second mutation that will let them continue to thrive. Yet this question is misguided, because there is no third sugar in the culture medium that another mutation could exploit.

The point here is as crucial as it is easy to miss. The mutation gave the microbes access to an existing supply of concentrated food;

it did not create the food out of thin air. If the complex sugar had not existed, the mutation would have accomplished nothing. As the complex sugar runs out, further mutations are possible — some microbes might feed on microbial wastes; others might kill and eat other microbes; still others might develop photosynthesis and make sugars from sunlight — but all these possibilities draw on resources much less concentrated and abundant than the complex sugar that made the first mutation such a spectacular success. No resource available to the microbes will let them flourish as they did during the heyday of the first mutation.

It's unpopular to suggest that human beings follow the same rules as microbes in a Petri dish. Still, the thought experiment parallels our current situation. The evolutionary leap that gave rise to the first technic societies some three hundred years ago did not create energy from thin air; instead, that leap made it possible for human beings to access concentrated energy that already existed. Now that those resources are running short, trying to replace them with even newer technologies is an exercise in futility. What is more, as this chapter will show, attempting to ramp up new technologies in the hope of providing energy for the industrial world is inherently self-defeating. A functional response to energy shortages must be found elsewhere.

the innovation fallacy

It's a common belief nowadays that technological innovation can always trump resource limits, but history shows otherwise. When the Second World War broke out in 1939, for instance, Nazi Germany had the most innovative technology on the planet; German scientists and engineers fielded jet aircraft, cruise missiles, ballistic missiles, smart bombs and many other advances years before anybody else.[1] Germany's great vulnerability was a severe shortage of petroleum, but here again the edge of technological innovation was on Germany's side. German scientists developed effective methods of coal to liquids (CTL) fuel production early on and the Third Reich put these methods to work on a massive scale as soon as it became clear that the oil fields of southern Russia were permanently out of reach.

The results are instructive. Despite large coal reserves and a massive effort to replace petroleum with CTL, the German war machine quite literally ran out of gas. By 1944 the Wehrmacht was struggling to find fuel even for essential operations. The Allied victory in the Battle of the Bulge in the winter of 1944–5 is credited by many military historians to the fact that the German forces did not have enough tank or aircraft fuel to follow up the initial success of their offensive.[2] The most innovative technology on Earth, backed up with abundant coal and an almost limitless supply of slave labor, proved unable to run a modern economy without access to petroleum.

Why did the German CTL initiative fail? There were a number of factors, but the crucial one is the difference between energy and *net energy*, which is the amount of energy produced from a resource minus the amount of energy that has to be invested in the production process. Energy resources vary dramatically in their net energy — for example, an oil well yielding light sweet crude oil under natural pressure can produce 200 times the energy it takes to drill the well, transport and process the oil, and get the products to their end use, while wind turbines produce six to eight times as much energy in their lifetime as went into making the turbine from raw materials and putting it to use. Coal's net energy varies drastically depending on the variety of coal, the method of mining and the depth the coal seam lies underground, but even under the best conditions coal yields a small fraction of the net energy of petroleum. Add the substantial energy cost of the CTL process, and the net energy of Nazi CTL fuel production was very low and may actually have been negative — that is, it may have taken more energy to produce coal-based liquid fuel than the fuel itself yielded when burnt.[3]

These same issues come to the fore in comparing today's energy crisis with the Western world's last two transitions from one primary energy source to another — the shift from wind and water power to coal in the late 18th century and the switch from coal to petroleum at the beginning of the 20th. In both cases, the new resource rose to dominate industrial economies while the older ones were still in use. The world was in no danger of running out of water power in 1750 when coal became the mainspring of the industrial revolution, and peak coal was still far in the future in 1910 when petroleum seized King Coal's throne.

In fact, it's inaccurate to say — as so many histories of energy do say — that coal replaced water power or that petroleum replaced coal. What happened instead is that coal was added to the existing fund of water power, and when petroleum became available, it was added to the total; since then, natural gas and nuclear power have been added as well. The amount of coal used today worldwide is larger than the amount used in 1910; for that matter, wood, the oldest of human fuels, is used as heavily today as it was in 1850.[4] This is one reason why replacing petroleum with some other existing fuel is not an option: all the other fuels are fully committed already.

This is not the way people in the 20th century thought of such things, of course. The fact that petroleum yielded more net energy than coal, and coal yielded more than waterwheels, made it look as though each new energy technology brought more abundant energy supplies. In the 1950s and 1960s many people expected nuclear power to do the same thing — readers who are old enough will recall the glowing images of atomic-powered cities of the future that filled the popular media in those days. Nothing of the kind happened, because nuclear power proved less economically viable than fossil fuels. Today nuclear power plants produce barely five percent of the world's energy supply, and that meager fraction has been achieved only because governments heavily subsidize nuclear technology for its military value.

The demise of the nuclear future echoes the failure of the German CTL program and also gives a good overview of the energy dilemma that faces the industrial world today. Pure uranium-238 contains a great deal of energy, but tons of ore have to be mined and processed to yield a very small amount of fuel and mining and processing costs energy. Building, operating and decommissioning a nuclear power plant costs even more energy, as does dealing with the spent fuel. The sheer complexity of the process makes the calculations fiendishly difficult but all told, the energy cost to mine, process, use and dispose of fissionable uranium probably uses more energy than the uranium itself yields.[5]

This points out the reason why fossil fuels are so important to today's industrial economies. Pound for pound or barrel for barrel, crude oil contains more energy than any other abundant naturally occurring substance on Earth, and it is much less costly to extract, process and use than almost anything else. Its net energy yield is therefore drastically higher than the alternatives. Other fossil fuels, though they fall short of petroleum's standard, still yield much more net energy than any of the other available resources. This is no accident; the fossil fuels are the end result of half a billion years of slow geological processes that compressed the stored sunlight of prehistoric life forms into highly concentrated forms. The other energy resources available to the modern world occur in much less concentrated forms, and renewables — especially sunlight and wind, the most popular candidates for tomorrow's power — are among the most diffuse of all. Since it takes energy to concentrate energy, the net energy of renewables is relatively low, and so trying to collect and concentrate them in order to power the extravagant needs of a modern industrial society is like trying to make a river flow uphill.

Current debates about using technological innovation to get us out of our energy predicament thus start from hopelessly flawed assumptions. The Earth's immense reserves of fossil carbon were essential to the birth of industrial civilization; certain gateway technologies allowed humanity to break into those reserves, but those technologies would have been useless unless the carbon reserves were there in the first place. Innovation is a necessary condition for the growth and survival of industrial society, but it is not a sufficient condition. If energy resources aren't available in sufficient quality and quantity, innovation can make a successful society but it won't make or maintain an industrial one.

Efficient use of the energy that can be harvested from sun, wind, water and living things is still the key to the ecotechnic future, but that key cannot be used effectively for many years to come. The same logic that made it economical to replace renewable resources with fossil fuels in the past remains applicable today. While fossil fuels are still available, their superior net energy yield means that they will almost always be more economical than energy from renewable sources. Furthermore, trying to build a new energy economy on the back of the existing one risks an unexpected economic backlash, due to an effect I have termed the paradox of production.

the paradox of production

This effect unfolds from the awkward fact that all alternative energy resources depend on an economy and an industrial system powered by fossil fuels, especially petroleum. Since oil is the world's cheapest and most concentrated energy source, once it was readily available it became the fuel of choice for nearly all the technologies that make other resources available. Discussions of energy costs thus miss the mark when they treat the prices of coal and oil, for example, as independent variables. The machinery used to mine coal, and the trains used to transport it, are powered by diesel fuel. When the price of diesel goes up the cost of coal mining goes up, and when supplies of diesel run short in coal-producing countries the supply of coal falters as well.

Every other energy source currently used in the industrial world thus gets a substantial "energy subsidy" from oil. To continue the example, oil contains about three times as much useful energy per unit weight as coal does and takes much less energy to extract from the ground, process and transport to the end user than coal does. Coal production benefits richly from these efficiencies. If coal had to be mined, processed and shipped using coal-burning equipment, all those efficiencies would be lost and a great deal of coal production would have to go to meet the energy costs of the coal industry. The same thing is true of every other energy source. The energy used in uranium mining, for example, comes from diesel rather than nuclear power, just as sunlight doesn't make solar panels. Today's industrial economies depend on petroleum's energy subsidy for much of the net energy surplus that makes industrialism possible. That dependence fuses with the challenges of declining oil production to boobytrap the future of energy.

Because of the ready availability of petroleum energy, a huge proportion of industrial society's capital plant — the collection of tools, artifacts, personnel, social systems, information and human geography, among other things, that provide the productive basis for society — has been designed and built to use petroleum-based fuels, and *only* petroleum-based fuels. Converting it to anything else involves much more than just providing another energy source. Think for a moment of the difficulties that would be involved in building the sort of hydrogen economy so often touted as the solution to the energy shortages of the near future.

For the sake of discussion, grant that the massive amounts of electricity needed to turn seawater into hydrogen gas in sufficient volume will turn up, despite the severe challenges facing every source proposed so far. Getting the electricity to make the hydrogen, though, is only the first of a series of tasks with huge price tags in money, energy, raw materials, labor and time. Hydrogen, after all, cannot be poured into the gas tank of a gasoline-powered car. It cannot be dispensed from today's gas pumps, or stored in the tanks at today's filling stations or shipped there by the pipelines and tanker trucks currently used to get gasoline and diesel fuel to the point of sale. Every motor vehicle on the roads, along with the rest of the infrastructure built up over a century to fuel those vehicles with petroleum products, would have to be replaced in order to substitute hydrogen for petroleum as a transport fuel — and much of this replacement would have to be done before the new system could take any significant amount of pressure off dwindling petroleum supplies.

The same challenge, in one form or another, faces every other potential replacement for petroleum. It's not enough to come up with a new source of energy. Unless that new source can be used just like petroleum, the petroleum-powered machines we use today will have to be replaced by machines using the new energy source. Unless the new energy source can be distributed through existing channels — whether that amounts to the pipelines and tanker trucks used to transport petroleum fuels today or some other established infrastructure, such as the electric power grid — a new distribution infrastructure will have to be built. Either task adds massive costs to the price tag for a new energy source; put both of them together — as in the case of hydrogen — and the costs of the new infrastructure will generally be much greater than the cost of bringing the new energy source online in the first place.

Factor the impact of declining oil production into this equation and the true scale of the challenge becomes clear. Building a hydrogen infrastructure — from power plants and hydrogen generation facilities through pipelines and distribution systems to hydrogen filling stations and hundreds of millions of hydrogen-powered cars and trucks — will, among many other things, take a very large amount of oil. Some of the oil will be used directly by construction equipment, trucks hauling parts to the new plants and the like. Much more will be used indirectly, since nearly every commodity and service in the industrial world today relies on petroleum in one way or another. Until a substantial fraction of the hydrogen system is in place, it won't be possible to use hydrogen to supplement dwindling petroleum production. Instead, the energy costs of building the hydrogen economy will create a massive additional source of demand, pushing fuel prices higher and making scarce fuel even less available for other uses.

The same thing is true of any other alternative energy system that attempts to replace an existing fossil fuel. The costs differ depending on how much of the existing infrastructure has to be replaced, but there's always a price tag — and nearly all of the energy needed to pay that price will have to come from fossil fuels, because those are the energy sources our civilization has on hand right now. If the new energy source can be produced and put to use with existing infrastructure, this effect may well be small enough to discount, but it is always there.

At this point the paradox of production can be neatly defined. If energy prices are high because supplies are limited, the obvious solution is to increase the supply by producing more energy. At the same time, if this requires replacing one energy resource with another that cannot be produced, distributed or consumed using the identical infrastructure, the immediate impact of such a replacement will be to raise energy prices, not lower them. The direct and indirect energy costs of building the new energy system become a source of additional demand that, intersecting with limited supply, drive prices up even further than they otherwise would rise.

If the new energy source turns out to be more abundant, more concentrated and more easily extracted than the source that it's replacing, this effect is temporary; if the new source can be distributed and used, at least at first, via old technology, the effect is minimized; if the new source is introduced a little at a time, in an economy reliant on many other sources of energy, the effect can easily be lost in the static of ordinary price fluctuations. All three of these were true of petroleum in its early days. It started as a replacement for whale oil as a lamp fuel, and thus was distributed and consumed in existing technology; decades later, it became a transportation fuel, and relied on the petroleum lamp-oil distribution system until a new one could be constructed on the basis of existing revenues; its other uses evolved gradually from there over more than half a century, until by 1950 it was the world's dominant energy source.

None of the proposed replacements for petroleum, though, have those advantages. None of them yield more than a small fraction as much net energy as crude oil under natural pressure and none have petroleum's unique mix of abundance, concentration, ease of production and distribution and fitness for a world of machinery designed and built for petroleum-based fuels. The fuel they need to replace is not simply a source of luxury lighting, but rather is the most important energy source in the entire world today. As we've seen, trying to solve the energy crisis by bringing new energy sources online will drive up the cost of petroleum and other existing energy sources further than they would rise on their own, since the direct and indirect energy costs of the new source and its infrastructure have to be met from existing sources. Pursued with enough misplaced enthusiasm, a crash program to bring some new energy source online in a hurry could drain enough energy, raw materials, labor and money out of a fragile system to drive it over the edge into collapse.

jevons' alternative

Fortunately, a proven way exists to counter the paradox of production, exert downward pressure on energy prices and free up resources for a constructive response to our predicament. This option, like the challenge it meets, unfolds from the difference between energy and net energy. Just as net receipts, rather than gross receipts, determine whether a business prospers or goes bankrupt, it's the net energy available to a society, rather than the total amount of energy it produces, that determines how much that society can do with its energy supplies. The energy costs that have to be factored into net energy, however, are not limited to *production costs* — the energy needed to turn a natural resource into energy ready for distribution. *System costs* — the energy needed to get the energy to its end use, whatever that happens to be, and use it — also have to be taken into account.

Most of the energy needed to produce a resource is expended over the useful life of the resource: a coal mine, for example, needs a certain amount of energy input in advance to get things started, but most of the energy input is expended day by day thereafter. Most of the system costs of a resource, by contrast, have to be put in before the resource can be used at all: before coal from a mine can power a toaster, for example, the railway that gets the coal from mine to power plant, the power plant itself, the power grid that gets electricity from the power plant to the toaster, and the toaster itself all have to be present, and all this takes energy.

Thus the system costs of an energy resource vary, sometimes dramatically, depending on the use to which the resource is put. Compare the net energy of using photovoltaic cells to power America's computers with the net energy of using photovoltaic cells to power America's automobiles. The production costs are the same in either case, but the system costs are totally different. Solar-powered computers use an existing distribution network (the electric power grid) and a mature technology (electronic computers), so the system costs are the same as any other way of powering computers. Putting the same energy to work powering automobiles requires building millions of new cars (if the electricity is used directly in electric cars), or a huge new network of fuel plants, pipelines and filling stations, in addition to millions of new cars (if the electricity is used to produce hydrogen), and all this new infrastructure has to be built and paid for before the solar transport system can be used. The massive initial investment needed to meet system costs for the latter option, in turn, makes the former a better bargain.

The recent ethanol boom in the United States is the poster child for the effect just outlined. In terms of production costs, ethanol made from American corn is a losing proposition: it takes more energy to provide the fertilizers, pesticides, tractor fuel, and other energy inputs to grow the corn and to ferment and distill it into fuel ethanol, than you get back from burning the ethanol.[6] The system costs of ethanol, on the other hand, are negligible: the US already has an extensive transportation system for getting bulk grains from farms to factories and existing distribution networks are perfectly capable of handling ethanol and ethanol-gasoline blends. Factories to turn corn into fuel were the only missing piece and misguided government grants and tax write-offs took care of that nicely. When political pressures to do something about energy became too strong to resist, ethanol — even with negative net energy — was the easy option, because it could be done with much lower initial costs than any of the alternatives.

The difference between production costs and system costs, however, can also be made to work in a helpful way. Since system costs as well as production costs have to be factored into net energy, it's possible to boost the effective net energy of a resource just as effectively by decreasing system costs needed to put it to use as by cutting production costs. Thus all potential uses of a given energy resource are not created equal; some have lower system costs than others, and thus have more net energy at the point of use. In a world facing significant declines in the availability of energy, it's therefore crucial to pay attention to system costs as well as production costs, keeping net energy as high as possible, so that systems necessary to community survival can be kept going more efficiently.

This approach relies on the logic of Jevons' paradox, which was first propounded by British economist William Stanley Jevons in his 1866 book *The Coal Question*. Jevons pointed out that when improvements in technology make it possible to use an energy resource more efficiently, getting more output from less input, the use of the resource tends to go up, not down. His argument is impeccable: as using the resource becomes more efficient, costs per unit go down, and so people can afford to use more of it; as efficiency goes up, it also becomes economically feasible to apply the energy resource to new uses, and so people have reason to use more of it.

Jevons' paradox has been used more than once to argue against conservation, on the grounds that using energy more efficiently will simply lower the cost of energy and encourage people to use more of it. This was quite true when the only thing constraining energy supply was price, but this is no longer the case; geological realities rather than market forces are placing hard limits on the upper end of petroleum production. Thus Jevons' paradox becomes a counterweight to rising energy prices: as energy costs rise, conservation and energy efficiency measures become more profitable and the decrease in systems costs helps hold the price of energy in check.

When energy supplies are limited, therefore, Jevons' paradox is an argument *for* conservation, not against it. During the 1970s, people across the industrial world downshifted their lifestyles and used energy more efficiently with relatively simple cost-effective measures. As a direct result, energy use dropped sharply — between 1978 and 1985, for example, world petroleum consumption went down by 15 percent. The measures that caused that sharp decrease cost much less than it would have cost to produce 15 percent more petroleum and could readily have been pushed further if the industrial world had not chosen instead to cash in the resulting gains and spend them on one final energy binge.

Conservation programs also have energy costs, of course. Unlike new energy systems, though, conservation projects can be carried out incrementally and their energy payback is immediate. A tube of caulk used to seal the gaps around leaky windows costs very little in energy terms, for example, and by the time the caulk is dry it is already decreasing the amount of energy wasted in home heating. Under most circumstances, furthermore, a dollar invested in conservation will save more energy than the same dollar invested in any form of energy generation will produce.[7] This point has been raised many times already in discussions of the future of energy, but it has yet to become — as it must eventually — the linchpin of a successful strategy for dealing with the end of the age of cheap energy.

The essence of that strategy is that we do not need more energy. Rather, we need much less energy than we use today, and the faster and more comprehensively we carry out a radical decrease in energy use, the easier it will be to make the transition to the future. So vast a percentage of energy used in the industrial world today is wasted on uses that do not actually benefit human beings in any meaningful way — for one example out of many, consider the land use decisions and social geography that require millions of Americans to spend an hour or more every day commuting to and from work — that a great deal of this decrease could be accomplished without placing an undue burden on anyone. The political difficulties implicit in pulling the plug on extravagant habits of energy, of course, preclude any chance that such policies will be deliberately pursued on a large scale in our lifetimes. Still, the volatile energy costs and spiraling shortages that will be so prominent a feature of the near and middle future can drive many of the same changes, if enough people take the time to be ready for them and thus demonstrate that a comfortable and humane lifestyle can be powered by a fraction of the energy used today in the industrial world.

Alternative energy technology has a place in such a strategy, but that place is a nuanced one. The same logic that shapes the economics of homemade raspberry jam, as discussed in Chapter Seven, applies equally well to energy: when energy is produced and used at home, without involving the wider economy of energy production and distribution, the gains to the family and community can be significant. Such energy technologies as passive solar space heating, solar hot water heating, homescale wind and hydroelectric turbines and the like thus have a significant place in any sufficiently broad conservation strategy. So does anything else that will allow human needs to be met without relying on outside power sources.

From these steps, in turn, the energy technologies of the ecotechnic future will evolve. Until fossil fuel subsidies no longer prop up the economy, there is no sure way to know which technologies will turn out to be viable in a human ecology powered entirely by renewable energy and which will not. The end of the energy subsidy will inevitably impose drastic changes in the availability of materials and the feasibility of manufacturing processes. For the time being, then, a dissensus-based process that encourages the widest possible range of experimentation on the local scale is probably the best step; as fossil fuel depletion proceeds, the new conditions that open up will make other steps possible.

master conservers

So far, the political leaders of the world's industrial nations have had very little to offer in response to the twilight of cheap energy. Most seem to think that the advice allegedly given to Victorian brides on their wedding nights—"Close your eyes and think of England"—counts as a proactive energy policy. Eventually they will have to think of a better response, if only because political survival does have its appeal.

Many of the resulting policies and programs will be counterproductive, and even more of them will be useless. In most of the nations of the industrial world, politics has long since devolved into a spoils system whereby different factions of the political class buy the loyalty of pressure groups among the electorate by a combination of ideological handwaving and unearned largesse. As long as that system remains in place—and it has proven enormously durable, surviving wars, revolutions and massive economic changes—most of the responses proposed to solve this or any other crisis will be aimed at pushing ideological agendas or rewarding voting blocs rather than actually doing anything constructive.

Still, it's by no means impossible that some positive changes might come out of government as the pressures of the emerging energy crisis take their toll. One resource ready to hand, though it is rarely recognized, has a great deal to offer. A project launched during the energy crises and resource shortages of the Seventies may help explore the best available response to the energy issues central to this chapter.

Most American states have agricultural extension services that provide farmers with new crop varieties and advice from agronomists. During the 1970s, a number of these services started Master Gardener programs to train and certify volunteers, who could then field gardening questions that would otherwise clutter up the agricultural extension's phone lines. These programs proved hugely successful and in many areas they have long since been woven into the fabric of community life. In the small Oregon town where I lived until recently, for example, a Master Gardeners booth sets up shop at the local farmer's market every Tuesday between March and November, staffed by a brace of volunteers who will happily help you figure out what's chewing on your cabbages or what soil amendments your blueberries need.

Around the same time, soaring garbage disposal costs led to the birth of another venture along the same lines, the Master Composter program. Less visible than the Master Gardeners, and usually funded by local waste disposal companies, the Master Composters have concentrated on teaching people how to set up backyard compost bins to take their yard waste and kitchen scraps out of the waste stream. Some city governments helped the Master Composters out by giving away free compost bins to anybody who attended classes; the reduction in garbage disposal costs was substantial enough that this made solid economic sense. A similar set of pressures led to the emergence of Master Recycler programs in other areas.

The late Seventies and early Eighties saw the birth and abandonment of another project of the same sort: the Master Conserver program, which ran pilot projects in Washington and Oregon states and apparently never caught on anywhere else. I was one of the participants, ranging from teenagers to retirees, who attended weekly classes in the auditorium of the old downtown branch of the Seattle Public Library. We studied everything from thermodynamics and caulking to storm window installation and passive solar retrofits. Those who completed the course took received their Master Conserver certificates after taking an exam and putting in volunteer work helping schools, churches, nonprofits and elderly and poor homeowners retrofit for energy conservation. I still have mine tucked away in a drawer, the way old soldiers I've known kept medals from the wars of their youth.

Such a program could easily enough be relaunched by local governments today. A quarter century of further experience with the Master Gardener and Master Composter programs on county and state levels would make it child's play to organize; the information isn't hard to find, and the dismal level of energy efficiency common in recently built houses and the like could make a Master Conserver program a very useful asset as energy prices climb and the human cost rises accordingly. For that matter, it is improbable that I am the only Master Conserver from those days who still has all the class handouts from the program in a battered three-ring binder or who keeps part of a bookshelf weighed down with classic conservation books — *The Integral Urban House, The Book of the New Alchemists, Rainbook,* and the like. I don't quite remember anybody in the last days of the program saying, "Keep your Whole Earth Catalogs, boys, the price of oil will rise again!" Still, the sentiment was there.

More generally, of course, the experiences of any of the 20th century's more difficult periods can be put to work constructively as we move deeper into the 21st century's first major crisis. The victory gardens and ingenious substitutions that kept the home front going during two world wars are well worth revisiting. So are the lessons learned from earlier crises. Still, the experiences of the Seventies offer a particularly rich resource in this regard. Close enough to the present to be part of living memory for many people faced with the same basic challenge of too little energy, too few resources and too much economic instability for an overheated and overextended industrial world, they parallel our present predicament too closely to be neglected.

One crucial lesson from that decade may be particularly worth keeping in mind. In the depths of the Seventies energy crisis, conventional wisdom had it that energy would just keep on getting more costly as a lasting Age of Scarcity dawned over the industrial world. That didn't happen and many of the plans made in those years fell flat when the period of crisis ended. I've suggested earlier that the end of the Industrial Age will most likely trace out a stepwise decline, with periods of crisis and breakdown punctuated by periods of partial recovery. This offers the hope of breathing spaces in which the lessons of each time of crisis can be assessed and put to use in dealing with the next. Those who respond to the energy crisis of our own time might keep this in mind when preparing for the complex future ahead.

Community

THE ROLE OF COMMUNITY in the deindustrial future has received a great deal of attention in recent books on peak oil, and for good reason. While a few voices still promote the survivalist strategy of holing up in a cabin in the woods with a stockpile of ammunition and canned goods until the rubble stops bouncing, most of those who try to scope out our future in advance have recognized that the community, not the individual, is the basic unit of human survival. Still, discussions have too often avoided the homely realities of the communities people actually inhabit today. They've fixated instead on bright visions of ideal communities on the borderlands of Utopia, on the one hand, and horror-story scenarios on the other.

Both these draw much more on current quirks of collective psychology than on the probable shape of the future. Consider the notion that mindless, marauding hordes will spill out of the inner cities in the industrial world's last days, ravaging everything in their path. This has been a popular nightmare for more than a century. Fictional treatments of the theme — Newton Thornburg's successful 1980 novel *Valhalla* is a good example — focus tightly on the color difference between urban invaders and suburban victims, making it clear that on this side of the Atlantic, at least, the fantasy of marauding hordes roots down into one of the lasting legacies of American racism, the terror of the dark Other on which the shadow of white America's unacknowledged desires has long been projected.

Similar unstated motives shape less obviously apocalyptic visions of ideal communities. It is common in many circles that speak of community in reverent tones, for instance, to neglect the actual communities in which they live, on the grounds that the towns and neighborhoods most Americans inhabit are not "real communities." A complex history shapes this habit. The people who talk most fondly of community in the abstract are often the children, and more often the grandchildren, of people who grew up in tightly knit communities and fled from them as fast as possible once military service, college or something else provided a way out. Life in community has its downside; the rose-colored glasses of the communitarian ideal make it easy to forget the petty politics and the tyranny of collective opinion that can make real communities stifling places to live.

The concept of relocalization — a shift from political, economic and cultural centralization to local autonomy, proposed as an urgent necessity by many peak oil activists — can foster a similar myopia. Proponents of relocalization are right that the end of cheap energy will impose potent decentralizing forces on modern nations. Many of them, however, present this as an unqualified good. Given that only a few decades have passed since white racists in the American South urged relocalization in defense of racial segregation policies, this may be questionable. There are still places in America, after all, where relocalization would promptly bring back racist "Jim Crow" laws, and many others where more currently popular scapegoats — gay people, religious minorities, illegal immigrants and the like — would face mob violence without the protection of Federal civil rights laws. Only in the eyes of those seeking Utopia is unfettered local power always a good thing.

I suspect that some awareness of these awkward realities remains even among the most vocal proponents of relocalization, if only because very few of them have abandoned the anonymous freedom of large cities and moved, say, to small Midwestern farm towns where a half dozen churches and the local Grange hall still anchor something like the close-knit communities of yesteryear. Some activists have broken away from the pursuit of ideal communities and focused instead on revitalizing the communities where they already live—a project as sensible as it is admirable, and one that will be explored further on in this chapter. By far the majority, though, have focused their efforts on lifeboat ecovillages.

lifeboat ecovillages

A lifeboat ecovillage, for those who haven't yet encountered the idea, is a new community that relies on as many as possible of today's green technologies, is built in a rural area and owns enough agricultural land and other resources to provide for its own essential needs so that it and its inhabitants can survive the collapse of industrial society. Many versions of the concept are in circulation these days; I will focus on a specific example here, since the problems with the lifeboat ecovillage idea only become visible once the practical details come into play.

The version of the lifeboat ecovillage concept I have in mind proposes to resettle much of America's population as soon as possible in villages of 5,000 to 10,000 people, compact enough that nobody will need to own or use a car. Each village will have an assortment of green technologies and a permaculture zone surrounding the village that can support the inhabitants on a diet of edible forest crops.[1] The basic idea is plausible enough; it seems quite likely, in fact, that a network of independent towns with populations in the 5,000 to 10,000 range might be well adapted to the human and natural environments of a deindustrialized world, though that's merely a guess at this stage. The process of getting there is the difficulty.

Work out the cost and the downside soon appears. Given a population of, say, 8,000 per village and an average of 4 persons per family, each community needs 2,000 new homes. If these homes are cheaper than the median US home—say, $250,000 apiece—the startup cost is $500 million for housing alone. Add to that the cost of community infrastructure—everything from water and electricity to a school, a library and the like — not to mention the forest land surrounding the village, and the price tag has approximately doubled to $1 billion.

Even if half the prospective residents already own their own homes and can pay for their new housing out of their equity — not a likely situation in the midst of today's real estate price collapse — and all the residents put in sweat equity in the form of unpaid labor, it's still going to cost a great deal. If 2,000 families were committed enough to the project to risk their financial future on it, it might be possible to make it happen. Still, the risk is huge and made even larger by the fact that the new village is going to have to provide jobs for all its adult residents — part of the point of the exercise is that nobody owns a car, remember, so commuting to the nearest city is not an option.

Nor can the village's inhabitants count on being suddenly transported to a deindustrial world where they can live by harvesting their forest crops and bartering skills among themselves. For many years to come, they will have bills to pay and national, state and local taxes as well. Will the new village be able to provide its residents with the jobs needed to cover these costs? Many towns of the same size are failing to do that right now. Behind the attractive image of a village in the countryside lies the hard reality of a billion dollar gamble for survival against long economic odds.

That billion dollar gamble, furthermore, would at best only take 8,000 people out of the automobile economy. Imagine a program that instead attempted to take ten percent of the US population out of the automobile economy, the smallest number that would be likely to have any real effect on the fate of industrial society. The price tag would be around $3.8 trillion in direct costs, plus the huge indirect costs involved in abandoning ten percent of the country's housing stock, residential and community infrastructure and so on. It would take generations for the savings in energy expenditures to make up for the huge initial outlay, and if the program turned out not to work — if, for whatever reason, the world on the far side of peak oil turns out to be unsuitable for this kind of ecovillage — all that outlay would have been wasted.

Not all lifeboat ecovillage proposals have so high a price tag, admittedly, but few are cheap, and none can be expanded very far without crushing financial impacts. The problems with projects of this sort, though, go beyond the financial dimension. I have met far too many people who don't know enough about plant care to keep a potted petunia alive and have never put in a full day of hard physical labor in their lives — most middle class Americans haven't — but talk enthusiastically about the life of subsistence farming they expect to lead in a lifeboat ecovillage as industrial civilization crashes into ruin around them. It's all very reminiscent of the aftermath of the Sixties, when a great many young people headed back to the land with equally high hopes. Most of them straggled back to the cities a few months or years later with their hopes in shreds, having discovered that fantasies of the good life in nature's lap make poor preparation for the hard work, discipline and relative poverty of life as a subsistence farmer.

Very few such experiments have yet appeared in the age of peak oil. Partly, of course, it's one thing to leave the city behind for a rural commune when you're nineteen years old and can put all your worldly goods into a knapsack, with room left over for dreams. It's quite another to do so when you're 40 and comfortable, with a family, a career, and retirement sufficiently close that the impact of your choices on your pension is always somewhere in your thoughts. Many more of today's peak oil activists resemble the second of these categories than the first, which goes a long way to explain the gap between the number of lifeboat ecovillages on the drawing boards and the number that have been built. There simply aren't many people who can abandon their modern lifestyles, help pay for a rural community and support themselves there for decades while the machinery of industrial society shudders to a halt around them.

Still, this is only one reflection of the broader problem mentioned above, which is that lifeboat ecovillages of the sort just outlined make no economic sense in today's world. However self-sufficient they may be in the future, they are not self-sufficient here and now, when they have to be built and paid for. Nor is it at all clear how soon this will change. In terms of succession, would-be builders of lifeboat ecovillages are like seedlings of some climax forest species trying to grow in a piece of land still covered with pioneer weeds. The conditions that would allow them to flourish have not arrived yet, and may not arrive for centuries.

For the foreseeable future, the challenge we face is that of navigating the transition to a world where lifeboat ecovillages might make sense; it will be up to our grandchildren's grandchildren to face the very different challenges of living in such a world when it arrives. The irony is that the communities where most of my readers live today — the towns and cities that dot the contemporary North American landscape — may turn out to be the survival communities so many people have been seeking somewhere else.

cities in the deindustrial future

This suggestion flies in the face of one of the shared assumptions of nearly every apocalyptic movement of modern times. Whether the imminent catastrophe du jour is nuclear war, pandemic disease, racial conflict, Communist takeover, fascist police state takeover, the Antichrist or what have you, the standard response is to propose a flight from the cities to some isolated location in the countryside. The core assumption underlying this is the belief that when the crisis finally arrives, cities will inevitably be deathtraps.

Those with a penchant for the history of ideas can trace this belief back to the Book of Genesis, where Lot flees from Sodom into the wilderness of Zoar just before the fire and brimstone hits, and other passages in the Old Testament that reflect the distrust of urban life the ancient Hebrews absorbed in their nomad years. That imagery played into an enduring American schism between the genteel urban society of the East Coast, with its gaze fixed on Europe, and the impoverished rural society of the hinterlands where a culture independent of European roots found its seedbed. Generations of circuit riders and revivalists played off the contrast between urban vices and rural virtues, simultaneously flattering their listeners, undercutting competition from denominations with European roots, and feeding on popular bigotries against Catholics and Jews at a time when most American members of these latter faiths lived in large East Coast cities.

By the middle of the 20th century, the same way of thinking helped drive the conviction that the best way to deal with the problems of urban America was to load up the moving van and leave the city behind for some comfortably middle-class suburb out of sight and reach of the poor. Thus it's not surprising that the same tune gets replayed in a different key in today's American secular apocalypses, which draw their audience mostly from the white middle class. Too many of the lifeboat communities imagined by today's peak oil writers are simply a projection of the fantasy of suburban escape onto a backdrop of apocalypse.

Step outside the cultural narratives that make a flight to rural isolation seem like the obvious response to peak oil, however, and things take on a very different shape. It's true, of course, that some cities are much too big and badly sited to survive the end of the age of cheap energy; tourists of the far future will likely stroll among the fallen skyscrapers of Phoenix or Las Vegas as their equivalents today visit Teotihuacan or Chaco Canyon. Equally, it's hard to imagine that downtown Manhattan or Chicago will become anything in the future but vast salvage yards, though the excellent harbors adjacent to these two cities will likely stay in use for millennia yet. Yet it's crucial to note that the vast majority of America's cities do not fall into either of these categories.

Imagine a city of between 20,000 and 200,000 people in an agricultural region; there are hundreds of them scattered across North America, so this should not be too difficult. In a sudden collapse, it could be a challenging place to be, but an overnight collapse is the least likely way for the downslope of the industrial age to play out.

In the far more plausible scenario of uneven decline and depopulation spread out over decades, such a city would have immense advantages over any lifeboat ecovillage. Located close to farmland, stocked with abundant raw materials in the form of buildings, cars and the like, and a large enough work force to allow division of labor and the production of specialty goods, the city could easily import food and other necessities by supplying trade goods to the countryside, the way cities in nonindustrial societies have always done.

These same factors make maintaining public order much less challenging. The rural brigandage that commonly springs up in the last years of civilizations could make life very difficult for a lifeboat ecovillage, but a city with an organized militia based on its police force and pre-decline National Guard units would be a much tougher nut to crack. Finally, even small cities have the cultural and social resources — libraries and colleges, community groups and local politics, among other things — to maintain civilized life even in hard times. In a deindustrializing world, all these things are potent sources of strength. While some will undoubtedly fail, cities may well be the most viable options for personal and cultural survival as the deindustrial age opens.

Historically speaking, the largely independent city-state surrounded by its own agricultural hinterland is one of the most common foundations for an urban-agrarian society, and societies that have attained broader geographical integration routinely fall back to the city-state pattern in times of trouble. Some variant of it is very likely across significant parts of North America in the deindustrial future. Some parts of the continent lack the agricultural and resource base to support such a pattern; others will likely be in the path of armed invasions or mass migration, in which case all bets are off. Elsewhere, though — especially east of the Mississippi and west of the Cascade and Sierra crests, where rainfall and soil quality make sustainable agriculture a good bet — urban centers are likely to play a significant role through the deindustrial Dark Ages and on into the successor cultures to come.

Preparations for this role, interestingly enough, are already taking place. More than a dozen US municipalities are already at work on their own peak oil contingency plans and many more are considering it. The seismic shift that has placed municipal and local governments out in front on several other issues, leaving national governments behind them in the dust, seems to be under way in the peak oil field as well. Several movements outside local government, furthermore, have begun to work along these lines to organize cities and towns for the world after peak oil.

The most widely known of them, and the most successful so far, is the Transition Town movement.[2] The core of this approach is that a small geographical area — a town, a village, an urban neighborhood or the like — can make the transition to a post-petroleum world by harnessing the ideas and efforts of local people. The plan, now available in book form, starts with a core of activists who raise public awareness, forge alliances with local interests and government bodies, put together a consensus vision for an attractive sustainable future and finally midwife the birth of a plan, modeled on that vision, that can be adopted by the community and put to work. It's an intriguing project. I have my concerns about the movement, centered on the difficulties of successfully imagining the future in enough detail to plan it in advance. The movement's reliance on consensus methods, which tend to create bland compromises based on conventional wisdom, also strikes me as a vulnerability. Still, the fact that a sustained effort to reshape cities in the interest of sustainability is being made is a very positive sign.

One way or another, though, many of the people looking for communities that can survive the end of the industrial age might want to consider the communities in which they already live or consider relocating to another community already on the map. Different cities have different things to offer; a regional center of 100,000 people has one set of resources and amenities, while a college town of 20,000 may have quite another, so there is plenty of room for differences of approach. This option may not have the romantic aura of the isolated lifeboat community, but will likely prove much more viable in the real world of the next few centuries.

the ecology of social change

More generally, the gap between romanticism and reality has become an unavoidable challenge in building communities at the end of the industrial age. An embarrassing number of proposals for creating new communities and social structures assume that the future can be better than the present in terms of every variable its authors consider relevant. The trade-offs and bitter choices that so often constrain real societies in the real world rarely find any echo in these comfortable plans. The inhabitants of these imagined communities do not have to choose between peace and freedom, between feeding the hungry and protecting the environment, or indeed between any two values; they can have it all.

Utopian fantasies built on this assumption are by no means harmless. The dangers show themselves with particular clarity in the contempt for democracy that is almost a badge of pride today among alternative thinkers. One recent growth industry has accordingly been the coining of new political systems that supposedly make representative democracy obsolete. These new systems are rarely all that new, and very few of them provide checks and balances to limit the power of those individuals with privileged roles. In some cases, in fact, rhetoric about freedom is little more than a fig leaf for straightforward ideologies of elite rule. One popular book along these lines, David Korten's *The Great Turning*, insists that certain people have reached a higher "developmental stage" than the rest of us and are thus naturally fitted to run the world.[3]

This rhetoric ought to be familiar enough to anyone who knows the history of the 1930s. Now as then, representative democracy is an easy target for its critics. Abuses of power and displays of incompetence happen in democracies and closed societies alike, but in democracies they are more likely to become public knowledge and can be denounced in comparative safety — those people who condemn today's democracies as "fascist," for example, can do so without any fear of being dragged from their beds in the middle of the night and hauled off to a prison camp. The messy realities of democratic politics, with its mixed motives and unequal distribution of power, provide opportunities for those who argue that democracy has failed utterly because it isn't perfect, and that any change must therefore be an improvement.

Despite centuries of hard evidence to the contrary, those arguments find eager audiences in difficult times. History is littered with the results. One lesson that needs to be learned from the last three centuries or so is that attempts at radical change backfire much more often than not. The successful efforts for change are usually those that pursue specific improvements or target specific injustices, while those that pursue grander agendas tend to fail the more completely and disastrously the more utopian their goals become. Nor are the results of more extreme projects always an improvement on even the least satisfactory status quo: from the tumbrils of the French Revolution to the killing fields of Khmer Rouge Cambodia, it has always been those radical movements that promised heaven on earth that yield the closest approximation to hell.

The consistent failure of efforts to create a perfect society can be understood in many ways, but one core issue too rarely addressed is that such attempts ignore central facts of ecological reality. Human societies are, after all, ecosystems; they evolve over many generations in relationship to other living things and the rest of nature, and establish complex balances and feedback loops that make the results of changes almost impossible to predict in advance. When human beings set out to re-engineer a nonhuman ecosystem to suit their own preferences, all too often they assume that their new ecosystem will remain stable, balanced and healthy *and* give them what they want. The results of this display of hubris are usually disastrous — and the same thing is commonly true of attempts to re-engineer human societies along similar lines.

The principles that shape the environmental relationships of other species and communities, after all, apply equally to our own species and communities. Like other living things, human beings depend for their survival on natural cycles and are subject to natural limits. Like communities of other living things, human communities — from villages to nations — are shaped by their history, adapt to their environments, face hard choices between competing needs and respond homeostatically in order to counter movements toward disruptive change.

Thus social change, like environmental change, needs to start from a clear sense of the limits in which human societies operate, whether those limits unfold from natural law, resource flows or the impact of history on the present. Accepting those limits means that trade-offs will have to be made between incompatible goals and scarce resources divided among competing demands. A public process for making those trade-offs collectively, and adjusting them continuously, would thus arguably be the best option for a society that attempts to grapple with the challenges of the long transition to the ecotechnic age.

Such systems exist, at least in rough outline, in the present political systems of the world's democratic nations. A strong case can thus be made for reforming and improving the political arrangements that have already evolved in existing societies, rather than trying to impose a rigid ideological gridwork on the organic realities of our collective life. The caution, compromise and room for necessary change that run through the American and Canadian constitutions and their equivalents elsewhere are there because their creators knew the power of political passions and the fallibility of institutions. That makes their handiwork all the more relevant in a future when caution, compromise and change will be desperately needed. These systems were also crafted to fit the measured pace of transport and communications in an age before fossil fuels; this suggests another way that the old pragmatic rules may turn out to be more relevant to the real world of tomorrow than the utopian charters of today's political rhetoric.

None of this denies the need for significant social change as the world moves through the next few centuries of succession. Still, attempts to reshape human society in the face of peak oil might best be pursued in a spirit very different from the one motivating the utopian ideologies of the present and recent past. Such efforts might begin with a public process to determine what changes are desirable, followed by an exploration of the ecological conditions that encourage desirable social change. Then, like tribal peoples who have learned to gently reshape their environment so that it fosters useful plants in place of noxious ones, those who seek change would work to bring about those conditions, keeping an eye on the results and letting experience rather than ideology guide their efforts. This would likely be a slow process, and not a glamorous one, but the payoff could be substantial in terms of constructive, lasting change.

As yet, such an art of applied human ecology or social ecotechnics has barely been imagined, and a great deal of effort and patience will be needed if it is to become a reality, but the attempt to better society by remaking it according to ideological models has failed so consistently that it's high time to try something else. Such an attempt, however, must start by confronting challenging issues surrounding the cultures of the industrial world — or, rather, what little remains of them.

Culture 11

AFTER DECADES OF "culture wars" in which all sides redefined the concept of culture to fit their own political objectives it's difficult to talk meaningfully about culture in the industrial world at all. It's doubly difficult in America after the systematic destruction of America's own national and regional cultures, their replacement with a manufactured pseudoculture crafted by (and thus mostly congenial to) America's urban intelligentsia, and the consequent revolt of many working-class Americans against the concept of culture altogether. Yet a conversation about culture needs to happen, because a living culture is among the most crucial tools any human community needs to face difficult and rapidly changing times.

Culture is memory. An authentic culture roots into the collective experience of a community's past and from this source draws meaning for the present and tools for the future. Thus culture is a constant negotiation between the living and the dead, as new conditions call for reinterpretation of past experience and redefine the meanings that are relevant and the tools that are useful. When a society gives up on these negotiations and abandons the link with its past, what remains is not originality but stasis, in which a persistent set of common assumptions and popular narratives are rediscovered and rehashed endlessly under a veneer of novelty. Even the most hackneyed notions can count on being described as new and innovative ideas unlike anything anyone has thought before.

The resulting confusions pervade contemporary debates about the future of industrial society. Consider the claim that industrial society will end suddenly, completely and soon. There is nothing new about this claim, which has been made regularly since the mid-19th century. There's rarely anything new in the latest arguments supporting the claim, either; most of them were already well-aged before such classics as Roberto Vacca's *The Coming Dark Age* dusted them off for a new audience in the 1970s. The theory of history in which societies rise over time to a peak of wealth, power and corruption, and then suffer total destruction, can be found in the Old Testament and underlies the religious rhetoric of apocalypse that coined most of the ideas being retailed by today's prophets of fast collapse.

The persistence of the sudden collapse theory in apocalyptic rhetoric, it has to be said, is not matched by a similar persistence in actual history. It's vanishingly rare for a society to collapse at the peak of its wealth and power, for the simple reason that wealth and power are two of the most effective means for staving off collapse. As a rhetorical reality, however, the abrupt collapse of unjust power has immense cultural resonance. People are lining up for the chance to say, "How art the mighty fallen!" over the corpse of industrialism. The irony is that most of them seem to think they thought of those words by themselves, for the very first time.

For another example of this sort of repetition, take the pronouncements discussed earlier that the current troubles of industrial society are the harbingers of an evolutionary breakthrough to a higher mode of being, where the problems that beset us today will have lost all relevance. Few claims about the future are so insistently described by their proponents as new and innovative thinking; even fewer have less right to that title. Glance through such classics of Victorian thought as Joseph Le Conte's widely read work *Evolution*, first published in 1888, and you'll find the same claims of imminent evolutionary transformation that fill so many popular books today.

The idea of an evolutionary breakthrough was necessarily a latecomer on the cultural scene, since a theory of evolution had to be invented first. Once Darwin took care of this detail, each generation has identified any crisis that makes the headlines as the birth pangs of the new humanity. The most intriguing detail about all this, again, is the way that an idea that's been rehashed more often than the average sitcom plot has been mistaken for a radically new concept.

A third example is the profusion of claims that everything will be all right if only the right people are given unchecked power. This sort of thinking has become unpleasantly common in some parts of the alternative scene, though definitions of the new ruling class generally center on abstractions such as Korten's "higher developmental stages," mentioned in Chapter 10, in place of more traditional categories such as race or wealth.[1] The arguments used to justify these schemes differ only in minor details from the ones used by defenders of aristocratic privilege in 19th century Europe. Since few of today's readers are familiar with these older literatures, though, a set of shopworn claims have once again been hailed as new and innovative thinking.

a failure of mimesis

As these examples suggest, the constant reappearance of the same new ideas has a troubling side. Many of those ideas have been tried repeatedly in the past and worked very poorly indeed. Despite their appeal, there is no reason to think that they will work any better in their latest incarnations. Thus it may be worth looking into the failure of cultural memory that gives these ideas the false glamour of novelty. In his scathing 1986 study of ideologies of gender in 19th century art, *Idols of Perversity*, Bram Dijkstra commented:

> In a world which stresses the value of individualism above all else, it is a primary requirement for the "self-confident" mind, to remain blind to the logical conjunction of personal ideas and the assumptions held by the "mass" of one's contemporaries. The ideas of "individual" thinkers, more often than not,

> are largely constructed from contemporary clichés. These clichés have merely been stripped of their baser trappings, of their rhetorical conventionality, in accordance with whatever happen to be the prevailing guidelines for the "individualistic ego."[2]

Step past Dijkstra's irritable prose and the point he makes is worth exploring. Today's faith in progress devalues the legacy of the past; it's symptomatic that one of the more crushing phrases in modern teen slang is, "Oh, that's all history." Without the depth perception that an awareness of the past brings, though, the only raw materials modern thinkers have to hand are clichés of popular thought, and their thinking simply takes the next logical step from the same starting points. Santayana's famous dictum may need revision: those who do not remember their history are condemned to rehash it, under the delusion that they are being original.

There's a way out of this paradox of unoriginal originality, though it's at least as paradoxical: the way to get genuinely new ideas is to deliberately learn and value old ones, so that the past is not simply rehashed in an unthinking way, but actively investigated as a source of insight into the present. Creativity, as Arthur Koestler pointed out many years ago, comes from the collision of incommensurable realities.[3] To put that in less lapidary prose, it's when the mind engages thoughtfully with two or more different ways of making sense of the same thing that it can leap to a new level of understanding and come up with something authentically new. Just as the 19th century collision between Western painting and the art of other cultures enabled the Impressionists to break through to new ways of seeing light and color, our chances of finding the new ideas we so desperately need will improve if the unstated assumptions of contemporary culture are highlighted by a deliberate exploration of the radically different ways of looking at the world provided by the past.

The failure of transmission that has turned the Western world's own past into a foreign country was explored in detail by Arnold

Toynbee in *A Study of History*, which put patterns of cultural growth and contraction at the center of the historical process. Toynbee's argument, insofar as it's possible to sum up thousands of pages of subtle reasoning in a few paragraphs, is that civilizations emerge when a creative minority inspires the rest of society with a vision of human possibility appealing enough to break through the "cake of custom," the body of tradition that shapes the lives of indigenous cultures.

The key to their success is mimesis, our incurable habit of trying to imitate what impresses us. Children who play at being superheroes — look, I'm Spiderman! — practice mimesis; adults who think about what they want to become, or what they want society to become, are doing the same thing. In indigenous societies, the models for mimesis are tribal elders and tribal traditions, which accounts for their immense stability. Civilizations rise, in Toynbee's formulation, when a creative minority with an openness to new visions becomes the focus of mimesis instead.

A civilization enters decline, in turn, when its dominant minority loses the ability to inspire and settles for the power to coerce. As its role as a source of inspiration dwindles, so does its role as the focus of mimesis. People stop wanting to become like the members of the dominant minority and aim their dreams elsewhere. This splits society into two unequal halves, the dominant minority clinging to power by ever more coercive means, and a subordinate majority that goes through the motions of participation but no longer shares its society's values and goals. The latter becomes what Toynbee called an "internal proletariat," expected to perform the work that maintains the civilization but deprived of its benefits and, as the schism in society unfolds, increasingly alienated from its values.

The internal proletariat has been deprived of its folk cultures by the destruction of traditional lifeways, and barred from participation in elite culture by class and income barriers that grow higher as the imperial stage proceeds. Finally the internal proletariat makes common cause with the external proletariat — the people of surrounding societies exploited by the civilization, who never had any stake in its survival to begin with — and everything comes crashing down. As the privileged classes find themselves stripped of wealth and power by the upwardly mobile warlords of the ensuing dark age, the imperial society's cultural resources no longer have any value in the eyes of the masses. The result is a feedback loop that amplifies the impact of collapse. Pious hands tore down the temples of Roman gods and recycled the mathematical papers of Archimedes to provide parchment for Christian homilies, for example, because most people in the post-classical world no longer felt any loyalty to the culture of their ancestors and turned to the creative minority of a rising culture for inspiration no longer available from what was left of the dominant minority of the old.

It's an intriguing analysis, and Toynbee was not averse to applying its lessons to the industrial societies of the 20th century in which he himself lived. In his view, the formerly creative minority of Western civilization was well on its way to becoming a dominant minority, maintaining its position by economic and political force, and the rest of his society was rapidly becoming an internal proletariat with no stake in the civilization of its rulers. He argued that the fault for this "schism in the body politic" lay squarely with the political classes of his time, who were increasingly unfit to lead, unable to follow and unwilling to get out of the way. The fact that Toynbee, a significant figure in British international affairs for much of the 20th century, was himself a member of the ruling minority he critiqued adds an interesting twist to his analysis.

The idea of mimesis can also be used to make sense of a great many other social phenomena in today's world. One theme that runs through many of today's peak oil discussions, for example, is the complaint that so few people are willing to do anything about the end of the age of cheap abundant energy. Even within the peak oil community, surprisingly few people have taken the simple practical steps that will make their own lives easier as energy becomes scarce and expensive — growing a vegetable garden, learning to get by on less energy and so on. Outside the peak oil community, very few people are listening at all.

From Toynbee's perspective, this is simply another failure of mimesis. Those of us who write and speak publicly about peak oil are trying to break through a "cake of custom" just as firmly entrenched as the traditions of any tribal society, but we have arguably been trying to do it with the wrong tools and in the wrong way. Denunciation does not do the job and neither do reasoned proofs and footnotes; both of these, entertaining as they are, quickly become exercises in preaching to the choir. It might be worth suggesting that new approaches are in order.

At the same time, the schism in society Toynbee sketched has broadened dramatically since his time. It bears remembering that in the nineteenth century, for example, opera counted as a popular entertainment medium and women of privileged classes practiced the same handicrafts as their poorer sisters. Nowadays very few such common factors connect, say, the upper middle classes of an East Coast suburb with the rural poor of a Midwestern farm state. Folk cultures have guttered out or survive only as museum pieces, while elite culture withdraws behind walls of obscurantism — compare the accessible and popular fine art of the late nineteenth century with the unwelcoming and often offensive product served up by today's art scene.

the twilight of culture

In a world lurching through economic crisis and the first wave of impacts from peak oil, it's easy to dismiss this implosion of culture as a minor issue, but such a dismissal is as much a symptom of cultural collapse as anything cited already. Culture is memory, and among the things it stores are the tools, insights and lifeways that served people in the days before our civilization started chasing the addictive rush of empire. Rome offers a useful example; by the time the Roman empire began coming apart and the grain ships no longer sailed from North African wheat fields to Italian wharves, nobody remembered how things had worked in the days when the classical world consisted of independent city-states producing most of their own necessities at home.

When civilizations fall, such losses are common. Each society, as it grows, evolves institutions to pass on its heritage; as those institutions crumble, new arrangements to preserve cultural heritage have to be invented or the heritage goes away. Sometimes those new arrangements prove sturdy enough for the job and cultural heritage makes it through the ensuing dark age to successor cultures on its far side. Sometimes the new arrangements prove too frail, the thread of transmission breaks and enigmatic ruins become the only legacy of a dead civilization.

Very few people nowadays have grappled with the possibility that the cultural heritage of modern civilizations might share this vulnerability. Partly this stems from a widespread failure to recognize just how much has been lost in the past. The twilight of the Roman world, for example, involved the loss of most of its literary, scientific and philosophical works; some of the greatest creations of Greek and Roman cultures survive only as fragmentary quotations in texts that happened to make it through the Dark Ages. In some branches of culture the loss was total; the Roman world's musical traditions were as rich and complex as any music in human history, yet all that survives of Roman music is one fragment of a single haunting melody that takes about 25 seconds to play.[4]

At a time when a few strokes on a keyboard can call up music from anywhere on Earth, a repeat of that experience may seem like the least of our worries. Today's technologies have made it possible to copy and store more information than ever before, and revolutions in information access has given more people access to cultural treasures than ever before. Sheer cultural overload seems more likely these days than cultural loss. Still, like every other aspect of contemporary life, the information technologies that underlie this access to cultural resources depend on cheap fossil fuels, and more broadly on the survival of complex technologies that will be unsustainable after the industrial age.[5]

The most obvious difficulty is that the forms now used to store information depend on specific technologies dependent on the industrial system. Data stored on Internet servers, for example, exists only while the servers get uninterrupted power and spare parts, and these can only be provided if industrialism remains viable. The same dependence affects most other data storage — the information on a CD, a roll of microfilm or a vinyl record can't be accessed without the right machine, and most of these machines use information-processing programs just as vulnerable to loss as the technologies themselves. Most of the data collected by NASA space missions in the 1960s, for example, sits uselessly in warehouses of reel-to-reel tape today, because the software needed to decode it no longer exists.

This difficulty is compounded by the vulnerability of the data itself. Most data storage media break down over very short time frames. The paper on which this book is printed, for example, will be brown and crumbling fifty years from now, and most other media have even shorter shelf life. A future in which the only trace left of Western music is one 25-second fragment of a single Bing Crosby tune may be closer than we think.

There are those who argue that this is a good thing. The same habits of idealizing the future and demonizing the past outlined elsewhere in this book have shaped contemporary attitudes toward culture. Plato's insistence that poets ought to be driven out of a perfect society finds its echoes in today's cultural discourse; plenty of voices on the left act as though the culture of the Western world exists only to provide them with something to rebel against, while their equivalents on the right give lip service to the same cultural history, while stretching and lopping it to fit the Procrustean bed of a radical religious ideology. These habits, and the political struggles that drove them, have shaped current ideas about culture in ways that make the loss of the modern world's heritage far more likely than it has to be.

All these forces, in their own fashion, were factors in the cultural losses that followed the decline of Rome. Still, the Roman world lacked the extraordinary sense of historical change that, as John Lukacs has pointed out, is one of industrial civilization's distinctive traits.[6] Roman writers in the twilight of their empire apparently never noticed that their experiences mirrored, say, the implosion of the Mycenaean world in the 13th century BCE, nor did Roman historians treat Rome's own past as a guide to the future. Thus it seems never to have occurred to the Romans of the late Empire that their civilization might need to be handed on to a very different future. The task of salvage was left to Irish monks, centuries later, and by the time they got to work a huge amount of material had vanished forever.

Our situation differs from theirs only because our sense of history makes it possible to place our own experience beside that of the Romans and other fallen civilizations and draw conclusions about the shape of our own future. We are arguably in much the same situation as the Romans of the late Empire. We have, as they had, an immense cultural heritage, nearly all of it vulnerable to the impacts of decline; at the same time we have done our level best to erase the heritage of local folk cultures at home and elsewhere, just as they did, and thus risk losing precious knowledge that might make it easier to weather the descent from today's imperial heights. The one difference is that it's possible to talk in these terms today and propose responses to what will be one of the most challenging features of the decline and fall of the industrial world.

cultural conservers

A few years back the American middle class indulged in one of the orgies of self-congratulation by which it periodically proclaims its opinion of its own historical importance. The inspiration for this particular outburst was a 2000 book entitled *Cultural Creatives* by Paul H. Ray and Sherry Ruth Anderson, which announced that the spread of certain fashionable ideas through the middle class meant nothing less than the imminent transformation of American society.

This was simply another example of the failure of cultural memory just discussed. None of the book's enthusiastic reviewers remembered that the same transformation had been announced just as confidently in Marilyn Ferguson's *The Aquarian Conspiracy* in 1980, Charles Reich's *The Greening of America* in 1970 and a long line of predecessors reaching into the early nineteenth century. Like so many of today's new ideas, in other words, this one has been around for a long time, just as the attitudes Ray and Anderson identified as hallmarks of their "cultural creatives" have been accepted by much of the American intelligentsia since the heyday of the Transcendentalists in the 1820s.

In a paradoxical way, however, the "cultural creatives" phenomenon goes beyond the endless rehashing of the same new ideas that dominates today's cultural discourse. Behind the rhetoric of innovation and originality was a very different reality: a sector of America's intelligentsia had discovered ideas their parents, grandparents and great-great-grandparents valued in their time, and applied those ideas to the present day. Most of the people involved in this rediscovery had no idea that they were doing this, and so never drew on the legacies of the Transcendentalists, the Theosophists, the Beat generation or any other expression of the same current of thought. Still, what they did unconsciously can be done in a more deliberate way.

What we may need most in our present predicament, in fact, is not "cultural creatives" but *cultural conservers* — individuals who take personal responsibility for preserving some part of the world's cultural heritage. That's a tall order, not least because the decline and fall of our industrial civilization will leave many of us scrambling for bare survival. Still, it's not an insurmountable challenge, and three themes provide the loom on which cultural conservers can weave the individual patterns of their own work.

The first of these is focus. The cultural heritage of the modern world is too vast for any one person even to encounter it all, much less to preserve significant elements of it in any meaningful way.

Thus each cultural conserver will need to choose one or, at most, a handful of traditions to conserve. The value of dissensus suggests that the best guide to the prospective cultural conserver in choosing a focus is sheer personal passion. The tradition that speaks to each person most deeply — be it tablet weaving or Wordsworth's poetry, mountain dulcimers or handbuilt radio technology, classical philosophy or the great American novels — is the one that will inspire that person to the efforts needed to pass it on to the future.

The second theme is simplicity. The more resources needed to maintain a cultural tradition, the less its chances of survival in a time of scarcity. Music that can be played on a handmade instrument is more likely to survive than music that requires a symphony orchestra and an opera company trained to exacting vocal standards. Some complex traditions can be stored in durable forms: the reasonings of Greek philosophers, for example, made it to the Renaissance because they were written down on parchment and stored in monastic libraries through the intervening centuries. In many cases it's possible to choose between simple and complex options for preserving a technology; a hand-operated letterpress, for example, is much easier to build and operate than a laser printer. Technologies that are less efficient in the abstract, as this example suggests, may be more durable in the deindustrial future.

The third is transmission. It takes more than one lifetime for a civilization to decline and fall, and so the flip side of preserving cultural heritage is the challenge of passing it on to younger generations. Those traditions that will have obvious economic value in an age of decline and disintegration have a huge head start here. It's improbable, for example, that today's intensive organic growing methods will be lost any time soon, since these provide a survival advantage to those who know and use them. Still, cultural transmission does not always follow the economic line of least resistance. Those who know must be prepared to teach, and use their knowledge in ways that meet community needs.

These three themes sketch a few rough lines on a very broad canvas. It's worth noting that many people have already taken on the sort of project I am outlining here, some quite consciously. Just as the "cultural creatives" could have benefited from placing their own projects in a historical context, a sense of what worked (and failed to work) in the past can help shape constructive responses to the challenges of cultural conservation. Alongside the dismal record of cultural loss during ages of decline, after all, history also shows that a motivated minority who take the long view can have a disproportionate impact on the survival of cultural heritage in hard times.

Consider the survival of the Jewish people and culture after the destruction of the Second Temple in 70 CE. Faced with the risk of cultural extinction, religious leaders drew on memories of the Babylonian captivity to launch one of history's most successful programs of cultural conservation. As rabbinic Judaism took shape, a large percentage of its traditions focused on preserving Jewish cultural continuity. "Why is this night different from all other nights?" asks the Passover ritual; the answer, freely interpreted, is that it embodies one of the distinctive historical experiences of the Jewish people, using potent tools of symbol and ceremony to counter the pressures toward assimilation.

Equally, the Catholic church after Rome's fall set in motion a massive salvage program that kept much of classical culture alive through the Dark Ages. Its motives differed from those that drove the founders of rabbinic Judaism; an expanding church needed clergy literate enough to read scripture, theology, and canon law, and this mandated the survival of the Latin literary culture that informed early Christian literature in the West. Thus generations of Christian schoolboys learned Latin prosody from Virgil, and enough of them acquired a taste for learning to launch the great age of Christian monastic scholarship and preserve countless cultural treasures for the future.

Other examples abound, from the Sanskrit academies of India to the bardic schools of early modern Scotland, but they share a crucial feature in common. For a cultural tradition to survive in an age of decline, it needs to find a constituency that values it enough to put the survival of the tradition ahead of more immediate needs. In traditional Judaism, keeping the commandments isn't something to file away for future reference whenever times get hard; it comes first, even ahead of personal survival. Similarly, the Benedictine monks who spent long hours copying manuscripts by hand in unheated scriptoria through the worst years of the Dark Ages could have led much easier lives outside their monasteries if the glory of God had not, in their eyes, outshone all the treasures of the world.

Thus the survival of cultural heritage must draw on emotional drives potent enough to override the tyranny of immediate needs and inspire the modest but unremitting daily efforts needed to keep traditions intact. This is especially true of the traditions of high culture, which often lack short-term survival value and require a sizeable investment of time and resources. It is above all true of modern culture, which has specialized in the mass production of information to such a degree that the ability to maintain adequate storage for all the knowledge our culture has amassed is already in doubt.

A willing constituency will be hard for any part of today's high culture to find, and without it, there is a minimal chance that anything more than fragments of that culture will reach the future. Still, there is a wild card in the deck, and its name is religion. Nearly all the classic examples of cultural conservation drew their motivation from religious beliefs. Is it possible that some of today's cultural heritage will find a home within the environs of a present or future religious movement?

religion and the survival of culture

Religion has taken the lead in preserving cultural heritage often enough that Arnold Toynbee made the concept a key theme in the later volumes of his *A Study of History*. In Toynbee's view, the fading years of a civilization form a seedbed for new religious movements. As decline continues, one or more of these movements becomes a major cultural force. As the civilization that nurtured it collapses completely the new religious presence fills the vacuum, salvaging what remains of the old civilization's heritage, and the concepts central to that religious vision become the framework on which a new civilization begins to take shape.

It's easy enough to see why religion should play so crucial a role in the transmission of culture. In a time of social disintegration, when institutions collapse and long-accepted values lose their meaning, only the most powerful motives can ensure that the work of maintaining cultural heritage will be carried on. Religion is the one force that has consistently been strong enough for the job. The same sense of transcendent value that leads martyrs to sing hymns as they are burnt alive can also inspire scribes to preserve and transmit knowledge to a future they will never see.

Some of the religions that have accomplished this work in the past have been old, while others were newly minted. In either case, though, the gap Toynbee traced between the dominant elite and the internal proletariat becomes a force to be reckoned with. Religion is among the things most affected by the schism in society, and so one of the classic signs of a society on its way to collapse is a widening religious schism along class lines. America offers an interesting example of this process in motion. As it entered its imperial phase around 1890, a significant minority of Americans broke away from the religious consensus of their culture — a consensus that used the forms of mainstream Protestantism but, in the name of the "social gospel," transformed that faith into a worship of moral progress in borrowed Christian dress.

The vehicle for this widening schism was Christian fundamentalism. Twice, however — in the 1920s and then again in the 1990s — fundamentalist leaders proved all too eager to cash in their ideals in exchange for crumbs of political power from the tables of the dominant minority. The result in the first case was a near-total implosion of the fundamentalist movement, and a repeat of that process seems increasingly likely today as fundamentalist churches move further away from their role as social critics to embrace unthinking partisan loyalties nicely calibrated to support the status quo. The failure of fundamentalism to establish itself as an alternative to the values of the dominant minority has opened the field to other religious movements, but none has yet succeeded in taking on the role Toynbee outlined.

At the same time, the mainstream Protestant-progressive religiosity of the dominant elite has widened into a consensus shared by most varieties of American Judaism, much of the English-speaking wing of the American Catholic church, and several forms of American Buddhism, not to mention a very large number of people who insist they follow no religion at all. That consensus took the prevailing belief in progress and mapped it onto the social sphere, redefining religion as a movement for personal and collective betterment rather than a way of contact with spiritual realities that transcend the human world entirely. What is often portrayed as a rising tide of tolerance among these traditions actually marks the embrace of a common ideology of social progress unrelated to the central commitments of the faiths in question, but easy to insert into the hollow shell of any religious tradition once awkward questions about transcendent values are quietly put on the shelf.

Thus it's hard to name a major religious movement in contemporary America, or for that matter most other parts of the industrial world, that is well placed to rise to the occasion as industrial civilization begins its decline and fall. At the same time, it's crucial to remember that we are still in an early stage of that process. A Roman scholar of 250 CE, say, who tried to guess at the religious forms that would rise to prominence during the empire's decline would have faced a ferocious challenge in sorting through the contenders. His world was awash in new religious movements, some homegrown and others from abroad; nothing special marked out the destinies of Christianity and Judaism from those of their competitors, and the religion that played the largest role in passing classical culture to the medieval world, Islam, didn't even exist yet.

In the same way, one of the religious movements that could pick up the remnants of modern culture and pass them on to the future might consist today of a few dozen people gathered around a charismatic teacher in a commune in Kentucky. Another might be a small sect founded fifty years ago in Brazil or Bangladesh that still awaits the brilliant missionary who will bring it to Europe or North America and transform it into a mass movement. A third might still be an inchoate current that will not find its prophet for another hundred years. The one thing safe to predict is that those movements will draw on the religious heritage of today's culture but reshape it in ways that will inevitably conflict with the conventional wisdom of our age.

New religious movements there will be, however, and it's far more likely than not that they will attract a growing number of followers as the industrial age stumbles toward its end. It's often said that there are no atheists in foxholes, and there tend to be few in times of decline. In ages when people believe their own efforts can bring them the goals their culture sets before them, it's rare for them to worry too much about the transcendent dimension of life; it's only when those goals become unreachable that the majority raise their eyes to other possibilities and, as Augustine of Hippo phrased it, perceive a difference between the City of Man and the City of God.

Efforts to link this religious impulse to the survival of today's cultural heritage will more likely succeed if those who make such efforts let go of the assumptions of contemporary culture, and make peace with religious forms even when they offend modern sensibilities. Thus there seems little hope in the suggestion that today's scientific thought ought to redefine itself as a religion for this purpose. The raw material of religion certainly exists in modern science, or rather *scientism*, the mythic ideology that has grown up around the simple but powerful logic of the scientific method. Carl Sagan, who did more than any other recent thinker to cast that belief system in religious terms, was arguably one of the significant theologians of the 20th century.

Scientism as it exists today, though, embodies the attitudes and values of the dominant minority as much as any of the more obviously religious forms mentioned above. From its long struggle to seize authority from religious institutions, scientism inherited a bitter hostility to explicitly religious belief systems. The no man's land between science and spirit forms perhaps the single most troublesome barrier to the survival of science in the deindustrial world of the future. Crossing that barrier, in order to transmit the modern world's greatest intellectual adventure to the future, poses one of the great challenges of our time.

Science 12

FROM TODAY'S PERSPECTIVE, the possibility that science may need saving is practically unthinkable. Scientific research plays a huge economic role in modern society, and science has also become a source of collective meaning that fills many of the roles occupied by religions at other times. Nowadays, when people wonder how the Earth came into being or speculate about the future of humanity, the resources to which most of them turn first are those of science. Scientific institutions have profited accordingly, building an immense network of universities, research institutes, foundations and publishers, subsidized by billions in government largesse.

Many of the same things, of course, could have been said about the priesthoods of Jupiter Optimus Maximus and his fellow gods in the glory days of the Roman Empire or the scribes of the Lowland Maya in the days before Tikal and Copán were swallowed by the jungle. Civilizations direct resources to their intellectual elites, because they can, and because the payoff in terms of their values is well worth the expenditure. The downside is that the intellectual heritage of a civilization typically becomes dependent both on the subsidies that support these institutions and on the ideological consensus that makes those subsidies make sense. In the decline and fall of a civilization, both the subsidies and the consensus are early casualties. Thereafter, the temples of Jupiter get torn apart to provide stones for churches, and the intricate planetary almanacs compiled by Mayan astrologers rot in the ruins of the temples where their authors once contemplated the heavens.

Project the same process onto our future and the vulnerabilities of science stand out clearly. Imagine a world forty years from now in which annual production of fossil fuels has dropped far enough that only the countries that produce them can afford to use them at all. Half the population in nations with fossil fuels, and 90 percent in those without, labor at subsistence agriculture; most of the remainder work in factories converting salvaged materials with hand tools. Public health has declined disastrously, poverty and hunger are pervasive and literacy rates are dropping steadily. Dozens of nations have simply collapsed, and populations are on the move as sea levels rise and rain belts shift. In America, the old canal network is being reopened by laborers with shovels, fuel shortages have crippled a rail network that never recovered from its 20th-century dilapidation and cars are a luxury for the very rich. Meanwhile army units struggle with guerrilla forces across the mountain West, while refugees from starving Japan ride the currents en masse toward North America's Pacific coastline.

In such a world, what role could modern science have? Certain branches of applied science, especially those applicable to energy and the military, will get funding as long as anything still exists to fund them. Most other applied fields will have to scrabble for scraps, though, while pure research will go begging because the spare resources simply won't be available to support it. Most of the facilities that make advanced research possible today will be boarded up and abandoned, if not stripped of raw materials by the government or looted by the poor. Outside of those working in a handful of institutions providing scientific advice to generals and bureaucrats, the only scientists would likely be those willing to carry out research on their own time and at their own expense

Significant science could still be done in such a future. Epochal scientific discoveries such as the theory of natural selection and Mendelian genetics were made with equipment that would be considered hopelessly inadequate for a middle school classroom today. The problem here is that the transformation of science from a pursuit of gifted amateurs to a professional activity backed by government and corporate funds was mostly complete a century ago. Today it would be hard to find many scientists who could pursue their research in a basement lab with homemade equipment. Thus science faces the same predicament as other elements of today's cultural heritage: it needs a constituency to carry it into the future or it may well vanish entirely.

saving science

This is not the way most people nowadays think of the future of science. Here, as elsewhere, the fallacy of independence distorts nearly every attempt to make sense of the future. If science and technology automatically trump any other factor, it makes sense to assume that those societies with the most science and technology will thrive no matter what else goes awry. Yet future societies in a world of scarce energy and resources will rely for survival on the more basic necessities of clean air and water, fertile soil and human labor. To the extent that preserving science and technology get in the way of these essentials, science and technology will sooner or later draw the short straw.

The awkward fact that must be grasped here is that science and its products are much less essential to human survival than most modern people assume. Used wisely and directed with more caution than has usually been given to it in recent times, science can supplement and strengthen the connections to nature on which human existence depends, but it cannot replace them. To the extent that today's scientific community ignores these issues, it may well find out the hard way that it is less indispensable than it thinks.

Keeping the scientific project going through a future of economic contraction and technological decline is thus a tall order. James Lovelock, one of the few scientists to grasp the vulnerability of today's science, has suggested creating a large book containing all current scientific beliefs about the universe—"the scientific equivalent of the Bible," in his phrase—that can be distributed widely in advance of our civilization's collapse.[1] I have argued elsewhere that this is perhaps the worst possible approach to saving science,[2] but Lovelock's suggestion at least raises a central question: exactly what are we trying to save?

That word "science," after all, includes a great many things. It's common to divide it by subject into disciplines such as biology, physics, chemistry and so on. Here, though, another division has more value. We need to consider science as *product*, science as *profession* and science as *process* to make sense of the predicament of science and craft a strategy for its survival.

Science as product is what Lovelock discussed: those facts and theoretical models about the universe currently accepted as true by the majority of scientists in relevant fields. Science as product is the easiest part to save, since a single book preserved in some dusty library could pass on a great deal of it, the way that Ptolemy's *Almagest* saved nearly the whole corpus of Greek astronomy. Just as the *Almagest* eventually became a millstone around the neck of later astronomers, though, science as product easily fossilizes into dogma. By treating science wholly as product, Lovelock's proposal risks reducing science to the rote repetition of doctrines accepted on the basis of blind faith; his description of his project as an equivalent to Holy Writ is not a good sign.

Science as profession is the system of trained personnel and infrastructure that keeps today's science going. This dimension of science is fatally vulnerable to decline for reasons already discussed: the economic troubles, political chaos and desperate exigencies of an age of decline are likely to shred the support system for today's science in short order. In a time when the destructive legacies of technology may loom larger than its fading benefits, too, the possibility of violent popular backlash against science and scientists cannot be dismissed. Unless some other constituency can be found to support science as profession through the difficult times ahead of us, its chances of survival are minimal.

That leaves science as process, or the scientific method: that elegantly simple fusion of practical logic and applied mathematics that was birthed in the 17th century and gave birth in turn to the modern world. This is the dimension that needs saving ahead of anything else, since without it science cannot be done at all. Ironically, it is also the most vulnerable of the three, since few laypeople have any exposure to it. Lovelock's dream of scientific Holy Writ simply reflects current reality: science as product has eclipsed science as process, so that people outside the scientific professions are taught to accept scientific doctrines on faith, rather than being encouraged to practice science themselves. If professional science faces extinction, there's a real chance that it could take the scientific method with it to its grave.

Some eloquent voices have argued that this might not be a bad thing. Writers such as Theodore Roszak and Lewis Mumford have pointed out that the practical benefits of science must be weighed in the balance against the dehumanizing effects of scientific reductionism and the horrific results of technology run amok in the service of ambition and greed.[3] Others have argued that scientific thinking, with its cult of objectivity and its rejection of human values, is fundamentally antihuman and antilife, and the gifts it has given us are thus analogous to the gewgaws Mephistopheles offers to Faust at the price of the latter's soul.[4]

These arguments make a strong case against scientism, the intellectual idolatry that treats science as a surrogate religion. I'm not convinced, though, that they make a case against the practice of science on the modest basis on which it has been carried on until recently: as a set of effective tools for making sense of the world around us. As the age of cheap abundant energy comes to an end, and the reach of our sciences and technologies scales back to fit the realities of life in a world of hard ecological limits, the overblown fantasies that encouraged people to make science carry their cravings for transcendence are likely to give way. If the scientific method survives the social consequences of that loss of faith, it could still bring immense benefits to future societies.

A useful comparison can be made between the scientific method and the greatest intellectual achievement of the civilization ancestral to ours, the logic devised by ancient Greek philosophers and systematized by Aristotle in the books of his *Organon*.[5] By the time Roman civilization began to come apart, logic formed the core of the dominant intellectual movement of the ancient world: the schools of classical philosophy. Its practitioners considered philosophy the highest attainment of human reason; most other people found it irrelevant, and this split between elite and popular cultures widened until it could no longer be bridged. When classical civilization finally fell apart, two of the main schools of late classical philosophy, the Stoics and Epicureans, went extinct while the third, the Neoplatonists, survived in truncated form, because the Christian church found Neoplatonist metaphysics useful for its theologians.

The end of classical philosophy was not, however, the end of logic. Parts of the *Organon* in Latin translation survived the fall of Rome, and all of it was translated into Arabic within a century or so of the Muslim conquests. The civilizations that rose out of Rome's wreckage thus had Greek logical method to hand from the beginning. Both the Muslim East and the Christian West made logic a cornerstone of their own intellectual cultures. When modern Western cultures first began to take shape, in turn, Aristotle's logic was still on hand and its influence pervaded the new ethos of science; it's hardly an accident that Francis Bacon titled his groundbreaking essay on scientific method *Novum Organum*, "the New Organon."

The scientific method could be handed to the future in the same way by some newer Organon. (For that matter, a readable translation of Aristotle's own *Organon* would be well worth passing to future societies, which will probably find logic just as useful as the founders of our civilization did.) Science could also be passed on in a more immediately useful form, though, by teaching the scientific method to people who have good reason to practice it straight through the twilight years of the industrial age. There are any number of ways that this could be done, but the one that likely offers the largest gains for the future draws on the ecological principles central to this book.

appropriate ecology

The challenge of saving science has been made much worse by the way scientific research has become intertwined with a sprawling and expensive system of institutions and technologies. That is not, strictly speaking, essential to the practice of science. Little science nowadays, however, is done outside this system, and few scientists have any experience carrying out research without its support. In a future when institutional science will have to compete against more immediate needs for a shrinking pool of resources, scientific research will need to find other ways to support itself, or it risks sharing the fate of the old Stoics and Epicureans.

A little-known 1981 article by biologist David Mason proposed an approach to environmental science that neatly escapes these difficulties.[6] The appropriate technology movement of the time, with its focus on technologies that could be built and maintained with local resources, provided him with the core concept of an *appropriate ecology*: a way of environmental research that uses readily available tools and resources in place of the complex technologies central to so much ecological research.

His point remains valid today. Very little equipment is needed to carry out a controlled test of soil amendments on garden plots, identify pollinators visiting an orchard or track turbidity and erosion along the banks of a local stream. Nor is it particularly hard to turn the knowledge gained in these ways to practical account: for example, using appropriate soil amendments to enrich garden soil, providing nesting boxes for bumblebees, seeding erosion-controlling plants in vulnerable places and the like. Even within the harsh limits of time and resources imposed by the curve of decline such projects would be easy to pursue, and their value to a society dependent on yearly harvests and stable ecological relationships would be obvious.

More ambitious projects could also be pursued along the same lines. A society that until recently has focused far too much of its attention on hedge funds, for example, might be wiser to direct some thought toward hedgerows. In traditional farming through much of western Europe, borders between fields were planted with a mix of shrub species which grew into dense hedges. Beyond their social function as a boundary impossible to miss and difficult to move, hedgerows serve as miniature nature sanctuaries providing habitat to beneficial insects, birds and small predators such as snakes and weasels, which help control agricultural pests. A hedgerow can be created with the simplest of tools, but knowing what shrubs attract the right mix of living things is a subtler matter. Once again, though, the payoff for such knowledge would be significant; as suggested in Chapter Eight, the ability to solve problems at the interface between human society and the natural world will likely become the basis for viable professions in the future.

Could this same logic be applied to other branches of today's science? In some cases, at least, the answer is almost certainly yes. The subculture of amateur radio discussed in Chapter Eight, for example, supports lively experimental work, much of it with homebuilt equipment, in electronics, radio astronomy and aeronomy — the study of the upper atmosphere, where the electrically charged ionospheric layers that reflect radio waves around the world form and disperse. From today's perspective, it's easy to compare these projects unfavorably to the kind of research that can be done with government grants and multimillion-dollar facilities, but such comparisons lose their force when the latter no longer exist and the remaining choice is between small-scale research and no research at all.

Those branches of science harder to practice in a low-tech environment will be harder, though not impossible, to preserve. One option is the creation of new social forms to provide support for the more expensive kinds of scientific research once today's scientific institutions become tomorrow's boarded-up buildings and another century's crumbling ruins.

The crucial point to recognize is that no special providence guards science. It has been said, and rightly, that nothing seems so permanent as an empire on the verge of collapse or so invulnerable as an army on the eve of total defeat. Like the shattered statue of Ozymandias in Shelley's poem of that name, the fragments of today's science could stand someday in a wasteland once filled with the cyclotrons and observatories of a vanished age. One task we face today is sorting through the many options to save those things most likely to be of value to the future. Significant parts of today's scientific knowledge certainly belong in that category.

toward ecosophy

At the same time, the critics of scientism mentioned earlier make points that cannot be dismissed out of hand. The core theme of their argument is that science makes a good servant but a very bad master. Few of them dispute the value of the scientific method as a way of answering questions about nature. It is when the questions that science can answer are treated as the only questions worth asking, and when the picture of the world that can be gained by scientific inquiry is mistaken for the world itself, that science begins to transform itself from a means of inquiry to an ideology destructive of all those human values that lie outside its scope.[7]

The shift from inquiry to ideology is a common occupational hazard in intellectual traditions. Too often the conditions under which the tradition's preferred tools can be used are mistaken for definitions of truth and the limits of the toolbox confused with the limits of reality. Classical logic followed the same course: the Greek and Roman philosophers took logic as their basic tool, studied all of reality that could be reduced to verbal statements and analyzed by the rules of logic, and consigned the rest to the *apeiron*, the realm of the formless and unknowable. The result was a series of brilliant but bloodless intellectual syntheses that debated the nature of justice endlessly, for example, without ever asking if there might be a moral problem with the classical world's economic dependence on slavery. The disconnect between the questions classical philosophy could answer and the questions people most wanted to ask finally became so intolerable that Christianity, which offered answers to those unasked questions, swept all before it.

The same process is already under way in the relationship between science and modern culture. What verbal statements were to classical logic, quantification is to the scientific method: any phenomenon that cannot be expressed in numbers cannot be investigated by scientific means. Recognizing these limits is not a condemnation of science but simply an acknowledgment of the fact that no tool is suited for every job. Yet many scientists have reacted to these limits by consigning anything that cannot be quantified and repeated to their own *apeiron*. Consider the number of psychologists down the years who have insisted that consciousness does not exist because it cannot be measured.[8]

Still, the natural tendency of a small child with a hammer to believe that everything is in need of a good pounding is not the only factor at work here. The inner life of emotions and meanings is among the things the scientific method handles poorly — it is very hard to quantify an emotion — and this blindness is particularly marked when it comes to the narratives that have gathered around science itself. Thomas Kuhn pointed out in his celebrated book *The Structure of Scientific Revolutions* that each branch of science rests on a set of paradigms that go unquestioned, and often unnoticed, except in those revolutionary periods when the gap between the paradigm and the evidence forces itself into view.[9] What he did not discuss, and only a few of the sociologists of science have explored, is the extent to which those paradigms unfold from exactly that nonrational sphere of human life which science itself analyzes so ineffectively.[10]

The scientific method, after all, is simply a set of practical tools for studying nature. Many of the claims made in its name cannot be defended by scientific means. No controlled double-blind experiment proves, or could possibly prove, that truths revealed by science are more important than those uncovered by any other means, much less that the scientific method is the best hope for humanity's future. The fact that many scientists have made these claims does not make them scientific. Rather, they are value judgments based on the ideology of scientism.

The same point can be made with even more force about humanity's supposed "conquest of nature." A military metaphor that defines humanity as Earth's enemy is an odd way to make sense of our relationship with the natural systems that sustain our lives. Still, scratch most common attitudes toward the natural world these days and the hackneyed image of Man the Conqueror of Nature is rarely far below the surface.

This is even true of many of the narratives central to modern environmentalism. When contemporary human cultures and technologies are placed by definition outside nature and necessarily opposed to it, and nature itself is portrayed in the sort of helpless bunny imagery satirized in Chapter One (as too often happens in environmentalist literature), arguments that attempt to critique the imagery of Man the Conqueror simply end up reinforcing it. Note how often current discourse about the environmental crisis fixates on the insistence that humanity has become so almighty that it can destroy the Earth and itself into the bargain. It's almost a parody of the old atheist gibe: to prove our own omnipotence, we have made a crisis so big not even we can lift it out of our way.

The conflict between these narratives and the hard realities of the predicament of industrial civilization could not be more stark. Human limits, not human power, define the situation we face today because the technological revolutions and economic boom times that most modern people take for granted were a product, not of science or such impressive intangibles as "the human spirit," but simply of a brief period of extravagance in which we squandered half a billion years of stored sunlight. The power we claimed, in other words, was never really ours, and we never "conquered" nature; instead, we raided as much of her carbon assets as we could reach and went on a spending spree three centuries long. Now the bills are coming due, the balance left in the account won't meet them and the remaining question is how much of what we bought with all that carbon will still be ours when nature's foreclosure proceedings finish with us.

Perspectives such as these are impossible to square with most contemporary attitudes about nature and humanity's place in it and conflict just as sharply with the Enlightenment faith in applied reason as the door to a better world. From the perspective of that faith, it is axiomatic that anything unsatisfactory is a problem in need of a solution and that a solution can be found. The suggestion that deeply unsatisfactory conditions cannot be solved but must be lived with, is at once unthinkable and offensive to a great many people. Yet if human life is subject to hard ecological limits, the narrative of human omnipotence falls, and a popular and passionately held conception of humanity's nature and destiny falls with it.

The critics of scientism have done valuable work in pointing out the flaws in that conception and the thinking that underlies it. Scientism drew its strength, however, from the natural human habit of using successful technique to define the universe. Hunting and gathering peoples have always seen the animals they hunted and the plants they gathered as the building blocks of their cosmos; farming cultures see their world in terms of soil, seed and the cycle of the year; the efforts of classical civilization to inhabit a wholly logical world and those of modern industrial civilization to build a wholly scientific one, are simply two more examples of the same pattern. Nor was scientism always as obviously maladaptive as it has become today. During the heyday of the industrial age, it directed human efforts toward what was then an enormously successful mode of human ecology. Scientism's faith in the limitless power of human reason turned out to be a case study in what the Greeks called *hubris*, the overweening pride of the doomed, but there was no way this could have been known in advance, and there is a valid sense in which scientism has become problematic today simply because its time of usefulness is over.

A strong case can nevertheless be made that the cultures best suited to the age ahead of us must take an attitude toward nature differing sharply from scientism — an attitude that starts from humility rather than hubris, remembering that "humility" shares the same root as "humus," the soil on which we all depend for our survival. That attitude offers few refuges for today's notions about humanity's place in nature. Still, just as Greek logic was pulled out of the wreck of the classical world and put to use in Islam and medieval Christianity, the scientific method — as well as a good deal of today's scientific knowledge, and even whatever remnants of today's scientific professions get through the next few rounds of crisis — could function just as well in a culture of environmental humility as they do in today's culture of environmental hubris.

The most likely bridge between the two is again the science of ecology. Its vision of nature has already inspired such impressive prefigurings of a culture of environmental humility as Aldo Leopold's land ethic and the passionate love of nature that fires so many of today's green activists. As just noted, every culture draws on the techniques it finds most useful for metaphors to make sense of the universe. The worldview of industrial civilization drew most of its fundamental ideas, and even more of its symbolism and emotional appeal, from the world that was revealed by Galileo and Newton in the seventeenth century, and embodied in the first wave of industrial technology a century later.

In the same way, the need for ecological knowledge in the wake of the industrial age makes the emergence of a broader way of thinking modeled on ecological science all but inevitable. Call that way of thinking *ecosophy*: the wisdom (*sophia*) of the home, as distinct from — though in no way opposed to — the "speaking of the home" that is ecology, or the "craft (*techne*) of the home" that is ecotechnics. Ecosophy exists today only in embryonic form and many minds and hands must contribute to it before it becomes a fully realized way of understanding the world. The work of the late Norwegian philosopher Arne Naess, who coined the term ecosophy, provides one starting point for this work; another will likely come from the work of Gregory Bateson, Ervin Laszlo and their fellow systems theoreticians, who fused cybernetics with natural science to reveal some of the unrecognized complexities of nature; still another might well unfold from the critics of scientism mentioned earlier in this chapter and from other figures, now on the intellectual fringes, who have leveled their own critiques at the hubris of modern thought.

One aspect of ecosophy, in turn, will unfold from a sense of history's relation to ecological process. This will likely take shape only when ecosophy has developed much further than it has today. Still, the shape of history is already a live issue. Beliefs about the meaning of history are central to the worldview of scientism — without belief in the possibility of endless progress and the implied promise of better times to come, for example, the industrial world's huge investments in basic research are hard to justify — and those beliefs have spread with the rise of scientism itself to become central to a great many of the ideologies of the present age.

Even as the industrial age draws to its end, therefore, many people rely on a faith in limitless progress for a sense of purpose and value in life, and so even the most tentative sketch of the future from the standpoint of human ecology needs to address the question of the meanings and shapes that history might have in the absence of progress. It's ironic that one of the best places to begin that discussion is to glance at a recent announcement — one of many down through the years — that history itself had come to an end.

PART III

POSSIBILITIES

The Ecotechnic Promise

IN RETROSPECT, 1989 may not have been a good year to announce that history was over. That spring, however, a US State Department official named Francis Fukuyama did just that in an article titled "The End of History?" Later expanded into book form, Fukuyama's claim got the fifteen minutes of fame Andy Warhol claimed everyone would receive in the future and sparked enough controversy in academic circles to justify a small bookshelf of discussions and rebuttals.[1]

Fukuyama's announcement is easy to misunderstand and even easier to satirize. He was not claiming, as many of his critics suggested, that what might more broadly be called historical events would stop happening. Rather, he defined history as a competition among systems of political economy, leading to the victory of the best; the collapse of the Soviet Union and the victory of the United States in his eyes marked the end of that competition. The triumph of "liberal democracy" — Fukuyama's term for the corporate-bureaucratic states that rule most of the world's industrial nations — proved, he argued, that it was the best possible system. Thus history is over. In the years to come, those states that have not yet adopted liberal democracy will do so and the world thereafter will bask in an endless afternoon of relative peace and prosperity, the closest approximation to Utopia that human nature allows.

Such claims had been made many times before. What made this one unexpected was that it came from a self-described conservative.

Not that long before "The End of History?" saw print, quotes from Hegel and portrayals of history as a movement in the direction of Utopia were the badges of the Marxist intellectual and were rejected by conservative thinkers. Fukuyama's article thus marked the end of the process by which the American right turned itself into a copy of the Marxist cadres it thought it was opposing. The fate of the neoconservative movement, in turn, echoed that of its Marxist equivalents: just as Marxist regimes claimed to ride the wave of the future, only to learn that every wave crests and then flows back out to sea, neoconservatives' bluster about their place as "history's actors" gave way to frantic excuses and finger-pointing against a background of military stalemate, political failure and economic catastrophe.[2]

The same interlacing of theoretical inevitability and practical failure can be traced in other movements that have claimed that history was on their side. Such movements have never been in short supply and there is certainly no shortage of them now. The claim that an evolutionary leap will take care of all of today's problems has already been discussed in this book. The so-called Extropian movement insists, along the same lines, that science will shortly know everything that matters, technology will gain unlimited powers and they will be able to upload themselves into robot bodies and go zooming off to the stars.[3] Religious teachings that announce the imminent arrival of a new world by divine fiat, neoprimitivist theories insisting that the downfall of industrial society will force humanity back to the hunter-gatherer lifestyle forever and many others all insist that the existing order of things is about to give way to a new world from which history in the ordinary sense will be banished forever.

The emotional appeal of these predictions cannot be doubted, but every such claim that has been tested by events has been flattened by the steamroller force of historical change. The learned doctors who pronounce history dead, and the poets, prophets and philosophers who compete with one another to write her epitaph, keep on being inconvenienced by the patient's awkward refusal to lie down and stop breathing. Yet a powerful current of ideas in contemporary culture drives the repeated insistence that history is about to end. Those ideas must be faced in order to complete this exploration of the twilight of industrial civilization and the slow emergence of the ecotechnic age to come.

history's arrow

This last approach to the ecotechnic future follows in the wake of one of the major intellectual conflicts of the last three centuries, the struggle over the philosophy of history. Expressed in these terms, the whole affair sounds like one of those tempests in an academic teapot that give scholars in unpopular fields something to do with their time, but that appearance misleads. Like the character in one of Molière's plays who was astonished to find that he had been speaking prose all his life, most people these days believe in a distinctive philosophy of history without ever quite noticing that fact.

This is hardly a new thing. One of the ironies of the history of ideas is the way that so many cultural themes that first surface in avant-garde intellectual circles are dismissed out of hand by the grandparents of those who will one day treat them as obvious facts. Modern nationalism, to cite one example out of many, began with the romantic visions of a few European poets, spilled out into the world largely through music and the arts, and only then turned into a massive political force that shredded the political maps of four continents. This is the intellectuals' revenge on an unreflective society: the men of affairs who treat the arts as amenities and dismiss philosophy as worthless abstraction spend their workdays unknowingly mouthing the words of dead philosophers and acting out poems they have not read on the stage of current events.

The modern philosophy of history has traced a similar trajectory over the years. The philosopher Karl Popper, who devoted much of his career to critiquing it, called this philosophy *historicism*.[4] It may be defined as the belief that history as a whole moves inevitably toward a desirable goal that can be known in advance. Exactly what goal history is supposed to have varies from one historicist to another; choose any point along the spectrum of cultural politics, and you can find a version of historicism that treats the popular ideals and moral concerns of that viewpoint as the destination of the historical process. All the details differ, but the basic assumption remains the same.

Behind the current popularity of historicism lies a complex history. The founder of the current of thought that gave rise to historicism was an Italian monk named Joachim of Flores, who lived from 1145 to 1202 and spent most of his life writing abstruse books on theology. Most Christian theologians before him accepted Augustine of Hippo's famous distinction between the City of God and the City of Man, and assigned all secular history to the latter category, one more transitory irrelevance to be set aside in the quest for salvation. Joachim's innovation was the claim that the plan of salvation works through secular history. He argued that all human history, secular as well as sacred, was divided into three ages, the age of Law under the Old Testament, the age of Love under the New, and the age of Liberty that was about to begin.

Some of his theories were formally condemned by church councils, but his core concept proved impossible to banish. Every generation of church reformers from the 13th century to the 18th seized on his ideas and claimed that their own arrival marked the coming of the age of Liberty. Every generation of church conservatives, in turn, stood Joachim on his head, insisted that the three ages marked the progressive loss of divine guidance, and portrayed the arrival of the latest crop of reformers as Satan's final offensive. As secular thought elbowed theology aside, Joachim's notion of history as the working out of a divine plan got reworked into secular theories of humanity's grand destiny.

Notable among these was the theory argued by the Marquis de Condorcet in *Sketch for a Historical Picture of the Progress of the Human Spirit* in 1794. A rich irony surrounds this work, because Condorcet was one of the French *philosophes* whose ideas lit the fuse of the French Revolution, and like so many of the intellectuals of his time, he believed that a better world could be planned out in advance and then put in place by the collective will. When the Paris mob toppled Louis XVI's feeble monarchy, though, what replaced it was not a happy republic of reason but a parade of tumbrils hauling victims to Madame Guillotine. Condorcet himself was among those condemned to death, and he wrote his *Sketch* while hiding from the revolutionary police. His unhappy experience did nothing to dent his faith in reason and progress; his treatise argued that human history was an inevitable rise from barbarism to a utopian future in which human life would undergo endless improvement.

Condorcet's vision of perpetual progress found many listeners, but an even more influential voice was already waiting in the wings: Georg Wilhelm Friedrich Hegel, who managed the rare feat of becoming both the most influential and the most unreadable philosopher of modern times. Hegel witnessed the turmoil that shook all of Europe after the French Revolution and came to believe that history functions through a series of conflicts between opposed forces. Each such conflict had a place in the grand process of historical change by pitting two equally valid ideals — thesis and antithesis — against one another and forcing a synthesis between them, leading to a final state of perfection in historical time.

Hegel's view of history became enormously influential, less through his own work — I challenge any of my readers to struggle through a chapter of Hegel's prose and come out the other end with anything but a headache — than through the writings of those he influenced. Political radicals at both ends of the spectrum pounced on Hegel's ideas before the ink was dry on the first edition of his *Philosophy of History*. Karl Marx used Hegelian ideas as the foundation for his philosophy of class warfare and Communist revolution, while Giovanni Gentile, the pet philosopher of Mussolini's Fascist regime in Italy, was also a strict Hegelian. For that matter, Francis Fukuyama, who played Gentile's role for the neoconservative movement, drew his theory of an end to history straight from Hegel.

Still, the spread of Hegel's ideas isn't limited to the radical fringes, or even to those who know who Hegel was. When peak oil comes up for discussion outside the activist community, one of the most common responses is, "Oh, they'll think of something." If the person who makes this comment takes the time to expand on it, the underlying reasoning typically runs like this: every time the world is about to run out of some resource needed for progress, "they" find something new, and the result is more progress. This is Hegel reframed in terms of economics: shortage is the thesis, ingenuity the antithesis and progress the synthesis; the insistence that the process is inevitable puts the icing on the Hegelian cake. More generally, the logic of historicism provides the core assumption behind "they'll think of something": history's arrow points in the direction of progress, and so whatever happens, the result will be more progress.

It may be a waste of breath to contend with belief systems as pervasive and deeply rooted as historicism, but the effort has to be made, if only because historicism has a dismally bad track record. Behind every historicist belief system that has been around more than a few years, back to Joachim of Flores himself, lies a trail of failed predictions of the imminent arrival of history's goal. (Joachim himself, for example, believed that the age of Liberty would arrive in 1260.) A methodology with a consistent history of false results is arguably not the best choice for making sense of our future.

Despite this history of repeated failure, it's understandable that attempts to force history along a line of advance toward some predetermined end would be popular. The alternative is to accept that history may be headed in no direction at all, and many people find this profoundly troubling. In Western intellectual history, however, the most influential alternative to historicism has made exactly this argument. That alternative, the theory of cyclic history, portrays the historical process as a turning wheel on which civilizations rise and fall, while the wheel itself remains fixed in place.

history's wheel

The opposition between cyclic and historicist theories of history goes deep. Both try to make sense of history, but they seek different kinds of meaning from it; they get different answers because they ask fundamentally different questions. At the heart of historicism is the intuition that history has a purpose, while the core of the cyclic vision is the intuition that history has a pattern — and "purpose" and "pattern" are by no means interchangeable terms. Most theories of cyclic history disavow the claim that history has an overall purpose, direction or goal; their premise is simply that certain patterns show themselves over and over again in the history of different societies, and these patterns can be studied to extract common principles from which, as in any branch of science, predictions can be made.

The intuition behind the cyclic theory has ancient roots. Most early civilizations believed that nations rose and fell in great cycles modeled on the turning heavens; the Sumerian priests who tried to predict politics from the stars, the Chinese historians who tracked the movement of the Mandate of Heaven from dynasty to dynasty, and the Mayan kings who picked days to go to war with their neighbors on the basis of an intricate calendar all made use of the common sense of cyclic time.[5] With the emergence of philosophy in ancient Greece, secular theories of historical cycles came into being, such as the *anacyclosis* of the Greek historian Polybius, which tracked city-states through a repeating sequence in which monarchy, oligarchy and democracy fall into decay and are replaced by the next in order.[6]

Secular theories of history had little currency in the Western world from the twilight years of the Roman Empire to the time of the Renaissance, though the Muslim world found room for them; the historical theory of ibn Khaldûn, cited in Chapter Five, embraced a cyclic view of the rise and fall of dynasties influenced by the Greek tradition. In the West, the Italian philosopher Giambattista Vico was the next great figure in the tradition of cyclic history. In some ways, he was the most influential writer ever to have tackled the subject, though his triumph came long after his own time. Vico, who lived from 1668 to 1774, spent his career in obscurity as an assistant professor of rhetoric at the University of Naples. His masterpiece, *Principles of a New Science Concerning the Common Nature of Nations* (usually trimmed down to *New Science*), was ignored for more than a century, but his ideas had an immense impact on 19th- and 20th-century thought. "Giambattista Vico," wrote historian Anthony Grafton, "bestrides the modern social sciences and humanities like a colossus"[7] — though, like too many other monuments of thought, Vico is more often quoted than read, and more often read than understood.

It has to be said that there are good reasons for this. Vico drew many of his ideas, and most of the forms in which he expressed them, from Renaissance thought; he began *New Science*, for example, with an ornate frontispiece designed to be used as a memory image, at a time when the Renaissance art of memory had dropped out of use throughout Europe.[8] What he called an "ideal eternal history" divided into Ages of Gods, Heroes and Men would probably have gotten more attention in his lifetime if he had abandoned these classic Renaissance tropes, and referred instead to a general theory of history divided into ages of Faith, Reason and Memory.

By an intriguing twist of fate, Vico's work became central to both competing schools of the philosophy of history. Hegel and Marx both drew heavily on his ideas while rejecting the cyclic history at their center, while every theory of historical cycles after Vico's time has borrowed from him. Arnold Toynbee, whose theories were discussed at some length in Chapter Eleven, was no exception. His sprawling twelve-volume *A Study of History*, which appeared in stages between 1934 and 1954, shows its debt to Vico throughout, though Toynbee pursued the study of historical cycles on a more comprehensive basis than ever before; he identified 20 civilizations that completed their life spans, along with seven others that had failed prematurely and one — Western civilization — that is still in process.

The most creative heir of Vico's work, though, was Oswald Spengler, who spent his career in a minor teaching post much like Vico's, and published *The Decline of the West*, his major work, in two volumes in 1917 and 1925. Spengler has been poorly treated by more recent historians; Joseph Tainter, for example, takes him to task in

The Collapse of Complex Societies for not providing a scientific account of societal collapse, which is a little like berating Michelangelo for not basing his Sistine Chapel frescoes on modern astrophysics. Spengler was not a scientist, and never pretended to be one; he was a philosopher of history, and tried to offer a clear philosophical account of the way that cultures rise and fall. His vision of history, however, drew so heavily on images and metaphors from the life sciences that his work approximates a proto-ecology of history.

At the center of Spengler's project was an exploration of the life cycle of cultures. A culture in his terms is an overall way of looking at the world, with its own distinct expressions in religious, philosophical, artistic and social terms. All of western Europe from roughly 1000 CE on, along with the European diaspora societies in the Americas and Australasia, belong to one culture, the Faustian. The classical world from Homeric Greece to the early Roman empire is another, the Apollonian culture; another is the Magian culture, which rose in Persia, absorbed the eastern Roman Empire, and survives as the Muslim civilization of the Middle East. Other Spenglerian cultures are the Egyptian, the Chinese, the Mesopotamian, the Hindu and the Mexican.

The rise and fall of a culture in Spengler's sense does not follow shifts in political or economic arrangements. It tracks the birth, flowering and death of a distinctive way of grasping the nature of existence, and everything that unfolds from that — which, in human terms, is everything that matters. The Apollonian culture, for example, embodied a unique vision of humanity and the world rooted in the experience of the Greek *polis*, the independent self-governing community in which everything was decided by social process. Greek theology envisioned a *polis* of gods, Greek physics a *polis* of elements, Greek ethics a *polis* of virtues and so on down the list of cultural creations. Spread around the Mediterranean basin first by Greek colonies, then by Alexander the Great, and finally by Rome, this way of seeing the world became the basis of one of the world's great civilizations.

That, according to Spengler, was also its epitaph. A culture, any culture, embodies a particular range of human possibility, and like everything else it suffers from the law of diminishing returns. Sooner or later, everything that can be done from within a culture — religious, political, philosophical, intellectual, artistic, social and so on — has been done, and the culture fossilizes into a civilization. Thereafter the same cultural forms repeat themselves like a broken record, and are abandoned once their emptiness becomes impossible to ignore. Traditions from other cultures get imported to fill the widening void and technology progresses in a mechanical forward lurch until the social structure finally crumbles beneath it. A civilization can endure for a long time, but most end when some less brittle society forces its way into the old civilization's territory and brings new possibilities with it. The Apollonian culture did this to ancient Egypt, the Magian culture to the Apollonian, and the Faustian culture to most of the societies of the world.

What made this prophecy of decline a live issue in Spengler's time was that he placed the twilight of Faustian culture and the beginning of its mummification in the decades around 1800. At that time, he argued, the vitality of the cultural forms that took shape in western Europe around 1000 began trickling away in earnest. Thereafter the Western world's religions fossilized into the repetition of older forms. Its art, music and literature lost their way; its political forms launched the fatal march toward gigantism that leads to empire and, in time, to empire's fall; and its science and technology continued blindly on their way, placing ever more gargantuan means in the service of ever more impoverished ends.

The predictions he made on the basis of this belief turned out to be remarkably prescient. He argued, for example, that the next major empire in Western history would be ruled by the United States; that the struggle between democratic societies ruled by business interests, on the one hand, and dictatorships that placed the economy under political control on the other, would be the key political theme of the 20th century; and that the traditional forms of

Western art and music would be abandoned in favor of new forms inspired by non-Western cultures. All these were startling claims in Spengler's time; all of them have become everyday realities.

history, meaning and choice

What makes Spengler's vision particularly useful in the light of the themes already discussed here, though, is the depth of his challenge to historicism. He argued that history can have no overall meaning, because it's impossible to talk of meaning at all except within the worldview of a given culture. Each culture evolves its own way of experiencing human life in the universe, and the only meaning humans can know is embodied in these distinctive worldviews. No culture's worldview is more or less true than any other, nor are the worldviews of cultures that arise later on in history an improvement in any sense on the ones that came before; each culture defines reality uniquely through its own dialogue with the inscrutable patterns of nature and human experience. Interestingly, Spengler applied this logic to his own work as well; he offered his theory not as an objective truth about historical cycles, but simply as the best account of historical cycles that could be given from within the perspective of Faustian culture.

When it got past superficialities, much of the criticism directed at Spengler's work over the last nine decades took aim squarely at the two claims just cited: his insistence that every culture's worldview is equally valid and that humanity therefore does not progress; and his suggestion that our Faustian culture has already finished its creative period and worked out the full range of its possibilities. What makes these claims unacceptable to so many people is precisely that they offend against the pervasive historicism of our age. Only the belief that history is headed somewhere in particular, with our civilization in the lead, makes his theses in any way problematic.

A less ethnocentric viewpoint makes these critiques difficult to uphold. To begin with, Spengler was right to point out that trying to rank worldviews of different cultures according to some scheme of progress yields self-serving nonsense. Ancient Egyptians understood the universe in one way, and modern Americans understand it in another, not because Americans are right and Egyptians were wrong — or vice versa! — but because the two cultures are talking about different things, in different symbolic languages. A worldview that describes the metaphysics of human life in the language of myth cannot be judged by the standards of a worldview that takes analysis of the physical world in the language of mathematics as its starting point.

To say that the industrial world's technological progress proves the superiority of its worldview merely begs the question, since the Egyptians did not value technological progress. They valued cultural stability and achieved it, maintaining cultural continuity for well over 3,000 years — a feat our own civilization is not likely to equal. By their standards, for that matter, our society's ephemeral fashions, cultural turmoil and incoherent metaphysics would have branded it as an abject failure at the most basic tasks of human social life.

Still, a strong case can be made that Spengler undervalued the process by which certain kinds of technique invented by one culture can enrich later cultures. A relevant example already discussed is classical logic, among the supreme achievements of the Apollonian culture, which was inherited in turn by the Indian, Magian and Faustian cultures. No two of these cultures did the same thing with that inheritance; a toolkit Greeks devised to pick apart spoken language was used in India to analyze the structures of consciousness, in the Middle East to contemplate the glories of God and in Europe and the European diaspora to unravel the mysteries of matter. Without Greek logic, though, some of the greatest creations of all three inheritor cultures — the rich philosophies of Hinduism and Buddhism, the great theological syntheses of Islam and Christianity or the fusion of logic with experience that gave rise to the modern scientific method — certainly could not have been done as easily, and quite possibly might not have happened at all.

This suggests that while history is not directional, it can be cumulative. Nothing in the history of cultures older than Greece suggests that the emergence of logic was inevitable, just as nothing in the subsequent history of logic justifies the claim that logic is developing toward a goal. Still, the toolkit of logic, absent before the Greeks, enriched a series of cultures that flourished after them. There are countless examples spanning the full range of human cultural creations; for a small but telling example, consider how the practice of counting prayers on a string of beads, originating in India, has spread through most of the world's religions. For another, consider the way that 40 centuries of East Asian agriculture inspired the organic growing methods that will likely be tomorrow's food supply. Every person who finds spiritual solace in prayer or meditation with a rosary, or has a backyard organic garden to help put food on the table, has reason to be grateful for the slow accumulation of technique over time.

Thus there is a fine irony in the insistence by so many people these days that evolution will both relieve us from having to deal with the consequences of our own mistakes and get history back on track to their imagined goal. They are right that the historical changes facing today's world are evolutionary in nature; their mistake lies in misunderstanding what evolution is. Cultures, like species, tend to collect those adaptations that meet their needs and discard the ones that don't. Thus those techniques that happen to meet the needs of many cultures tend to survive more often than those that do not, just as those cultures that are able to make use of a suitable range of inherited techniques are more likely to thrive than those that do not.

This does not mean that the accumulation of useful techniques is the meaning, purpose or goal of history. Meanings, purposes and goals do not exist objectively as part of the brute facts of existence; they exist only when they are created and applied creatively by conscious persons, and a shared belief in meanings, purposes and goals by more than one person depends on the relation between the person proposing these things and those who choose to accept or reject them. (Atheists may read this statement in one sense, and religious people in quite another; interestingly enough, the logic works either way.)

Like biological evolution, furthermore, cultural evolution is in no sense inevitable. The crises that surround the decline and fall of civilizations, in particular, very often become massive choke points at which many valuable things are lost. One valid response to the crisis ahead of us thus could be a deliberate effort to help the legacy of the present reach the waiting hands of the future. The same logic that leads the ecologically literate to do what they can to keep threatened species alive through the twilight of the industrial age, so that biological evolution has as wide a palette of raw materials as possible in the age that follows, applies just as well to cultural evolution.

Thus it may not be out of place to imagine a list of endangered knowledge to go along with today's list of endangered species, and to take similar steps to preserve both. There are certainly other meanings, purposes and goals that can be found in, or more precisely applied to, either the inkblot patterns of history as a whole or the specific challenges we face right now, in the early stages of industrial civilization's decline and fall. We can decide as individuals whether to build on the heritage of our culture, to explore the legacies that have been handed down to us from other cultures, or to scrap the lot and try to break new ground, knowing all the while that other individuals will make their own choices, and that the relative success of the results, rather than any preference of ours, will determine which choices play the largest role in shaping the future.

The second of Spengler's controversial claims needs to be understood in the same nuanced way as the first. His claim that Faustian culture finished its creative age in the 19th century is a generalization, as any broad statement about history must be; as generalizations go, however, it has a great deal of insight to offer. Consider the arts: those that have their roots in the Faustian world, if they are still practiced at all, have either fossilized into repetitions of old forms, like classical music; turned for inspiration to the arts of other cultures, like popular music, which draws heavily from African music by way of the influence of blues on rock and jazz on nearly every contemporary genre; or become the self-referential concern of a narrowing circle of cognoscenti, like today's avant-garde art music. Similar patterns can be traced straight across the spectrum of Western cultural forms. What Spengler suggests is that this is not blameworthy; rather, it is the natural fulfillment of the cultural life form that sprouted in western Europe around 1000, burst into flower at the beginning of the Renaissance, and has now gone to seed.

The botanical metaphor is one that Spengler himself would have appreciated, but it points to the same undervalued dimension of historical process discussed earlier: the way that the discoveries and creations of a culture can be transmitted across time. Spengler's view of what he called civilization — the second half of a culture's life cycle, when its creative possibilities have all been worked out — was largely negative. The ancient Egyptians would have disagreed strenuously. From their viewpoint, geared to cultural stability and the preservation of traditional forms, what a Faustian mind necessarily sees as a creative period becomes a matter of groping in the dark, and what a Faustian mind sees as stagnation is the healthy form of a successful society. Nor can the Egyptian viewpoint be dismissed out of hand. Maintaining living traditions of cultural continuity, a rich and tolerant religious life, and stunningly beautiful art and architecture for more than three thousand years is not a small achievement.

Even within a Faustian perspective, the fulfillment of the Western world's cultural trajectory has potentials worth recognizing. To return to a familiar example, the sorting process that picked Aristotle's *Organon* out from among scores of other Greek works on logic, and spread it throughout the Mediterranean world, happened long after the creative age of Greek philosophy was over. As culture gives way to civilization, a ruthless winnowing of cultural heritage begins, and those creative works and techniques that survive the process become basic to the arts, crafts and sciences of the mature society. From there, along routes already explored in this chapter, they move past the periphery of that particular civilization and become part of the common cultural heritage of humankind.

This is the phase in which Spengler's vision placed the Western world of his time. Whether his scheme makes sense of the broader phenomenon of historical change he hoped to clarify, it provides a perspective crucial to our own time. The end of the age of cheap energy has many implications, but one of the most important — and most daunting — is that it marks the end of the road for nearly all the cultural trends that have guided the industrial world since the revolutions of the 18th century. Those trends pursued greater size, greater speed, greater power, replacing human capacities with ever more intricate machines, demanding ever more abundant energy and resource inputs, and escaping from the interdependence of living nature for an artificial world transparent to the human mind and obedient to the human will.

That way to the future is no longer open. The nations of the industrial world could pursue it as far as they did only because a wealth of fossil fuels and other natural resources were available to power Faustian culture along its trajectory. The waning of those reserves and, more broadly, the collision between the pursuit of unlimited economic growth and the hard limits of a finite planet, mark the end of those dreams. They may also mark the beginning of a time in which we can sort through the results of the last three centuries, discard the ones that worked poorly or demand conditions that no longer exist, and keep what still has value. The question that remains is how the historical process itself might best be understood in this context.

the eyes of nature

The ecological perspectives central to this book have rarely been applied to the philosophy of history. For that matter, the borderlands between ecology and philosophy — one of the places from which a future ecosophy will most likely arise — have been studied so far only by a handful of thinkers.[9] Still, points raised in earlier chapters offer some sense of the way human ecology becomes history.

The first and most necessary of these steps, it seems to me, is the recognition that history is an ecological phenomenon, governed by the same laws as other processes in nature, and cannot be understood at all when approached from a purely anthropocentric standpoint. Our species is finally beginning to learn that treating the environment as though it exists solely for humanity is a self-defeating habit, and the same recognition needs to be extended to history as well. Human history is only partly a function of human choices and deeds; at least as important is the role of nonhuman nature, which plays a major role in all historical processes and a central role in many.

Once history is recognized as an ecological phenomenon, the next step is to search history for processes that appear across the range of ecosystems in the nonhuman world, and look for their equivalents in human affairs. I have suggested in this book that three such patterns can be traced in the human past and show every evidence of continuing to shape the human future. The first of these is the rhythm of rise and fall that shapes the history of civilizations, as traced by the theories of cyclic history. It parallels population cycles throughout nonhuman ecosystems and, like those cycles, allows for some degree of prediction: when a human society expands beyond the limits of its environment, for example, the arrival of crisis followed by steep decline is rarely far off. In human terms, this pattern works on a time scale of centuries.

The second such pattern is the succession process that replaces R-selected social forms with a series of more K-selected forms, ending in the social equivalent of a climax community that remains stable until changes in the environment disrupt it. If there is a linear aspect to history, this pattern provides it, though succession has very little in common with the historicist theories discussed earlier in this chapter: succession moves toward stability, not toward Utopia, and even when stability is achieved—which does not always happen — it is never permanent, because the environment is always subject to change. Within these limits, succession allows some degree of prediction: when a human community relates to its environment in unsustainable ways, for example, a gradual shift toward more sustainable patterns will usually take place thereafter. In human terms, this pattern works on a timescale of millennia.

The third such pattern is the process of cultural evolution that gradually accumulates useful techniques and every so often leverages one or more of these, by way of some previously unused resource base, into a sudden leap into a new form of human ecology. This pattern is neither circular nor linear; instead, like biological evolution, it branches outward along whatever lines of advance may be available, and the breakthroughs that define new human ecologies come at unpredictable and widely separated intervals. Unlike historical cycles and succession, evolution offers no help to prediction at all; fewer than half a dozen major evolutionary leaps have occurred in the history of our species, though there have been many more minor ones, and the timing of those leaps appears random even in retrospect. In human terms, this pattern works on the scale of deep time, defined by the lifespan of the human species.

The historical setting of modern industrial civilization can only be grasped in full against the background of all three of these patterns. In the first or cyclic pattern, as my earlier book *The Long Descent* showed, industrial civilization is in the early stages of its decline and fall. The central claim of that book is that a significant number of the events — more exactly, the *kinds* of events — that will happen to industrial societies over the next one to three centuries can be known in advance by paying close attention to equivalent phases in the histories of other civilizations. From an ecological perspective, this argument remains valid, but it is not the whole story.

In terms of the second or succession pattern, industrial civilization is a profoundly R-selected form of technic society, dependent on linear processes that turn resources to waste and undercut the ecological base on which the system itself depends. As they decline and fall, today's industrial societies will thus be replaced by more K-selected social seres, which will be replaced in their turn as succession continues. Our civilization's place in succession tells us little about the kind of events to expect in the foreseeable future, but it provides two other insights: first, an idea of the direction in which a wide range of future changes will tend; and second, some gauge of how far those changes will likely proceed with the emergence of the next historical cycle, the one I have called scarcity industrialism.

In terms of the third or evolutionary pattern, the irony that surrounds claims that an evolutionary leap will solve industrial civilization's problems can be fully appreciated. The leap these claims predict, as Chapter Four pointed out, has already happened; it took place three centuries ago, with the birth of industrial civilization. Far from solving our world's problems, it is directly responsible for them, having pitched us headlong into a world of new possibilities and dangers unknown to previous human ecologies.

The challenges placed in our path by our species' latest evolutionary leap, in turn, will be solved only through a long and difficult process of trial and error that slowly replaces the unsustainable habits of today's industrial society with the sustainable lifeways of the ecotechnic societies of the far future. Our civilization's place as the first tentative sketch of a technic society tells us little about the events that will shape this trajectory into the future, and not much more about the trends that will define and direct those events. What it provides instead is something potentially more valuable, at least for a species like ours that thrives on meaning: a sense of how we got into our present fix and how we might begin to extract ourselves from it.

This kind of meaning differs in important ways from the historicist claim that it is possible to know the purpose of history or, for that matter, from the simplistic sort of cyclic theory that insists on a rigid sequence of stages around the wheel of time. The meaning offered by evolution provides no final answers and places no limits on human freedom, for good or ill. If today's industrial civilization is the first of many technic societies, that fact (if it turns out to be a fact) says nothing about where human social evolution might proceed. Some new evolutionary leap might happen later on, if adaptation to circumstances places human communities in a situation that opens up new possibilities, but there can be no guarantees that this will take place. For that matter, the coming of the ecotechnic age is a probability rather than a certainty. Common ecological patterns point that way, but again, there are no guarantees.

To see humanity's trajectory into the future in this way, ultimately, is to consider ourselves through nature's eyes rather than our own: to see ourselves as one species among many, uniquely gifted in some ways but far from unique in others, and subject to the same natural laws and ecological patterns as every other living thing on Earth. Troubling though this view may be to the hubris cultivated by some recent human cultures, it reaches back to the older and more ecologically relevant ways of understanding humanity and nature common to traditional cultures. If some of the more speculative sections of this book guess right, this view may also reach forward to a future ecosophy, and thus to ways of making sense of the world that will blossom in the far future, as the ecotechnic age dawns.

Afterword

THE WIDENING GAP between the future most people in the industrial world once expected and the one they seem most likely to encounter remains poorly grasped even now, when the 21st century we were supposed to get is rapidly becoming a subject for satire. Still, an awareness of the troubling changes now cutting the ground out from under today's lingering faith in progress has taken root in many corners of the world. What was merely a vague current of unease not so many years ago shows signs of taking on a social role of no small significance. In the near future, as the failure of progress becomes harder to miss, the need for new visions of the future may become an overwhelming force.

How that force will shape our collective life in the decades to come is still anyone's guess. Some of the proposals offered nowadays to cope with the crisis of the industrial world may well offer promise, while others just as popular today embody deeply troubling commitments to the utopian politics I have tried to critique in the preceding pages. No one knows yet which of those projects will go forward to shape the future and which will be remembered only as might-have-beens. In the broad sweep of historical and evolutionary time, at least, these are early days yet.

Early or not, decisions we make here and now may well have a significant impact on the way the rest of the process plays out. Much of the knowledge, many of the skills and a fair selection of the tools that will likely be necessities further down the slope of decline are readily available right now. The crises of the present and recent past have closed a few doors but left many others open; the industrial world's libraries have not yet been gutted, some of its schools still offer an education worth having and the Internet, for all its dubious features, still provides a medium for worldwide discussion and information sharing. None of these things will be around by the time the decline of the industrial age is well under way and the vagaries of history make it impossible to guess how soon any one of them may vanish, so acting sooner rather than later may be the wisest course.

Those who take up the challenge suggested by this book and make the effort to address the needs of an age not yet born can choose among many directions for their work. The possibilities sketched out in Part Two of this book suggest some of the available options. Those who want a clearly defined plan of collective action can adopt one from any of the several movements and organizations in the peak oil community that have drawn up such plans. I have my doubts about each of these efforts, but the logic of dissensus applies to my own ideas just as much as anyone's; there is always the chance that one or more of these schemes might work.

Still, the broader picture is worth keeping in mind. Our species has launched itself on a journey through time far more complex and less predictable than anything either the historicist or cyclic theories of history can encompass. The road to the ecotechnic future can only be guessed at in advance, and will have to be built step by step as the human societies of the future struggle to adapt the legacies of our age to the hard limits of a finite planet and the unguessable possibilities of their own time. What we do now, or leave undone, may have a potent influence on their successes or failures. Challenging though it will certainly be to take action on that basis, I can think of no task more richly worth our efforts.

Notes

Introduction

1. See especially William R. Catton Jr., *Overshoot: The Ecological Basis of Evolutionary Change*, University of Illinois Press, 1982; and Richard Heinberg, *The Party's Over: Oil, War, and the Fate of Industrial Society*, New Society Publishers, 2003; and Donella Meadows et al., *The Limits to Growth*, Universe, 1972.
2. See, for example, Julian Darley, *High Noon for Natural Gas: The New Energy Crisis*, Chelsea Green, 2004; Kenneth Deffeyes, *Hubbert's Peak: The Impending World Oil Shortage*, Princeton University Press, 2003; and Heinberg, *The Party's Over*.

Chapter One: Beyond the Limits

1. The two paragraphs that follow are based on the discussion of energy and nutrient cycles in Eugene P. Odum, *Fundamentals of Ecology*, W.B. Saunders, 1971.
2. The distinction between cyclic and linear ecosystems given here is based on Catton, *Overshoot*.
3. Fossil fuel consumption figures are from the *BP Statistical Review of World Energy 2008*, available from bp.com.
4. World annual energy consumption equals around 500 exajoules (one exajoule = 10^{20} watts per second), with some 86% of this coming from fossil fuels. Total solar energy absorbed by green plants annually is estimated at 2000 exajoules by Akihiko Ito and Takehisa Oikawa, "Global mapping of terrestrial primary productivity and light-use efficiency with a process-based model," in M. Shiyomi et al., eds., *Global Environmental Change in the Ocean and on Land*, Terrapub, 2004.
5. See the *BP Statistical Review of World Energy* for each of the last ten years, available from bp.com
6. Compare, for example, current energy debates with the discussions of America's energy future in Lane deMoll, ed., *Rainbook: Resources for Appropriate Technology*, Shocken, 1977.

7. This is the central theme of Meadows et al., *The Limits to Growth*; see also Robert L. Hirsch, Roger Bezdek and Robert Wendling, *Peaking of World Oil Production: Impacts, Mitigation, and Risk Management*, US Department of Energy, 2005.
8. R. Costanza et al.,"The value of the world's ecosystem services and natural capital," *Nature* 387 (1997), pp. 253–260.
9. C. Wright Mills, *The Causes of World War Three*, Simon and Schuster, 1958.
10. See Ray Kurzweil, *The Singularity is Near*, Viking, 2005, for an egregious example.
11. See"How civilizations fall: A theory of catabolic collapse," in John Michael Greer, *The Long Descent: A User's Guide to the End of the Industrial Age*, New Society Publishers, 2008, pp. 225–240.

Chapter Two: The Way of Succession

1. See the many examples cited in Clive Ponting, *A Green History of the World: The Environment and the Collapse of Great Civilizations*, St. Martin's, 1992.
2. Jared Diamond, *Collapse: How Societies Choose to Fail or Succeed*, Penguin, 2005, is the classic ecological analysis.
3. Odum, *Fundamentals of Ecology*, provides the model of succession on which this section is based.
4. This way of approaching the history of agriculture differs sharply, of course, from the version common in alternative circles these days, which interprets the invention of agriculture as a form of "original sin" — sometimes quite literally; see, for example, Daniel Quinn, *Ishmael*, Bantam, 1992. See Colin Tudge, *Neanderthals, Bandits, and Farmers: The Origins of Agriculture*, Yale University Press, 1998, for a survey of recent (and less polemical) scholarship on the origins of agriculture, on which this section is based.
5. Ernest Callenbach, *Ecotopia*, Banyan Tress, 1975, is the classic example.

Chapter Three: A Short History of the Future

1. Nassim Nicholas Taleb, *The Black Swan: The Impact of the Highly Improbable*, Random House, 2007.
2. See, for example, the archives of housingpanic.blogspot.com, where nearly every element of 2008's financial crisis was discussed at length up to three years in advance.
3. See John Kenneth Galbraith, *The Great Crash 1929*, Houghton Mifflin, 1954, for a discussion of the repetitive and predictable nature of speculative booms and busts.
4. See the account in Myron J. Echenberg, *Plague Ports: The Global Urban Impact of Bubonic Plague, 1894–1901*, New York University Press, 2007.

5. In Catton, *Overshoot.*
6. See Deffeyes, *Hubbert's Peak,* for a discussion of the Hubbert curve.
7. Intergovernmental Panel on Climate Change, *Climate Change 2007*, Cambridge University Press, 2007.
8. See Corale L. Brierley et al., *Coal: Research and Development to Support National Energies Policy*, National Academies Press, 2007.
9. The most drastic of these shifts, around 11,500 years ago, sent global temperatures rising 12°C in less than 50 years. See sciencedaily.com/releases/2008/06/080619142112.htm.
10. See Ponting, *A Green History of the World*, and Richardson B. Gill, *The Great Maya Droughts: Water, Life, and Death*, University of New Mexico Press, 2000.
11. Thomas Friedman, *The World is Flat: A Brief History of the Twenty-first Century*, Farrar, Strauss and Giroux, 2005.

Chapter Four: Toward the Ecotechnic Age

1. See, for instance, David Korten, *The Great Turning: From Empire to Earth Community*, Berrett-Koehler 2006; Alexia Parks, *Rapid Evolution: Seven Words That Will Change Your Life Forever*, Education Exchange, 2002; and Graeme Taylor, *Evolution's Edge: The Coming Collapse and Transformation of Our World*, New Society Publishers, 2008.
2. Compare Bram Dijkstra, *Idols of Perversity: Fantasies of Feminine Evil in Fin-de-Siècle Culture*, 1986, and Martin Fichman, *Evolutionary Theory and Victorian Culture*, Humanity Books, 2002.
3. The discussion of bat evolution given here is based on R.L. Carroll, *Vertebrate Paleontology and Evolution*, W.H. Freeman, 1988, and Nancy B. Simmons, "Taking wing: Uncovering the evolutionary origins of bats," *Scientific American*, December 2008 (online edition, sciam.com).
4. See discussion of evolution's nonlinear nature in Stephen Jay Gould, *Wonderful Life: The Burgess Shale and the Nature of History*, W.W. Norton, 1990.
5. See Stephen Jay Gould, *Time's Arrow, Time's Cycle: Myth and Metaphor in the Discovery of Geological Time*, Harvard University Press, 1987, pp. 1–8.
6. The idea that hunting and gathering societies are by definition free of war, social inequality and similar troubles has been promoted enthusiastically in recent years but cannot be supported by the evidence. Many hunting and gathering tribes among North America's First Nations, to cite one example out of many, engaged in fierce tribal warfare and slave trading. See, for one example out of hundreds, V.F. Kotschar, *Fighting with Property: A Study of Kwakiutl Potlatching and Warfare, 1792–1930*, University of Washington Press, 1950.

7. Good introductions are James A. Beckford, *New Religious Movements and Rapid Social Change*, SAGE Publications, 1986; and Anthony F. C. Wallace, *Revitalization Movements*, Bobbs-Merrill, 1956.
8. Alice Beck Keyhoe, *The Ghost Dance: Ethnohistory and Revitalization*, Holt, Rinehart and Winston, 1989, is a sympathetic history.
9. Barbara W. Tuchman, *The Proud Tower: A Portrait of the World Before the War, 1890–1914*, Macmillan, 1962, offers a good summary of the world before 1914.
10. See Dmitry Orlov, *Reinventing Collapse: The Soviet Example and American Prospects*, New Society Publishers, 2008.
11. This is discussed in Alf Hornborg, *The Power of the Machine: Global Inequalities of Economy, Technology, and Environment*, Alta Mira Press, 2001.
12. For useful studies of neoconservative ideology and agendas see Mark Gerson, ed., *The Essential Neoconservative Reader*, Perseus, 1997; and Jacob Heilbrun, *They Knew They Were Right: The Rise of the Neocons*, Doubleday, 2008.
13. See Howard Odum, *Environmental Accounting: Emergy and Environmental Decision Making*, 1996 for emergy accounting.

Chapter Five: Preparations

1. See Marie Louise von Franz, "The Process of Individuation," in C. G. Jung, ed., *Man and his Symbols*, Doubleday, 1964, for this concept.
2. C. G. Jung, "Wotan," in C. G. Jung, *Civilization in Transition*, Princeton University Press, 1970.
3. See John Kelly, *The Great Mortality: An Intimate History of the Black Death, the Most Devastating Plague of All Time*, HarperCollins, 2005.
4. Diamond, *Collapse*, again provides the analysis used here.
5. See ibn Khaldûn, *The Muqaddimah: An Introduction to History*, tr. Franz Rosenthal, Princeton University Press, 1967.
6. Ewa Plonowska Ziarek, *An Ethics of Dissensus: Postmodernity, Feminism, and the Politics of Radical Democracy*, Stanford University Press, 2001.

Chapter Six: Food

1. Several million acres each year complete the conversion process from conventional to organic agriculture; see Economic Research Service, United States Department of Agriculture, "Briefing room: Organic farming and marketing," ers.usda.gov/Briefing/Organic
2. Documented in David Duhon, *One Circle: How to Grow a Complete Diet in Less Than 1,000 Square Feet*, Ecology Action, 1985; and John Jeavons, *How To Grow More Vegetables Than You Ever Thought Possible on Less Land Than You Can Imagine*, Ten Speed, 1979.

3. See Willy Schilthuis, *Biodynamic Agriculture*, Floris, 1994; and Rudolf Steiner, *Agriculture: An Introductory Reader*, Biodynamic Agriculture Association, 1958.
4. For introductions to permaculture see Graham Bell, *The Permaculture Garden*, Thorsons, 1994; Bill Mollison and David Holmgren, *Permaculture One*, Transworld, 1978; Bill Mollison, *Permaculture Two*, International Tree Corps Institute, 1979; and Patrick Whitefield, *Permaculture in a Nutshell*, Permanent Publications, 1997.
5. Orlov, *Reinventing Collapse*, offers a good outline.
6. Stu Campbell, *Let it Rot!: The Gardener's Guide to Compost*, Storey Publishing, 1990, one of the standard handbooks, is my source for much of this section.
7. See Joseph Jenkins, *The Humanure Handbook: A Guide to Composting Human Manure*, Chelsea Green, 1999.
8. Carol Steinfeld, *Liquid Gold: The Lore and Logic of Using Urine to Grow Plants*, Ecowaters, 2007.
9. The "hygiene hypothesis" remains controversial but is supported by epidemiological evidence. See David Strachan, "Hay fever, hygiene, and household size," *British Medical Journal* 299 (1989), pp. 1259–1260.
10. See Stu Campbell, *The Mulch Book: A Complete Guide for Gardeners*, Storey Publishing, 1991.
11. Recent disputes around the ethics of eating animal foods are complex and, in my view, badly in need of clear reasoning — enough so that limits of space do not permit a detailed discussion here. It may be clear from the following, however, that I find claims of the immorality of eating animals unconvincing.

Chapter Seven: Home

1. In Le Corbusier, *Towards a New Architecture*, Dover, 1986.
2. See, for example, the introduction to Frank Lloyd Wright, *Truth Against the World: Frank Lloyd Wright Speaks for an Organic Architecture*, Wiley-Interscience, 1987.
3. In Stewart Brand, *How Buildings Learn: What Happens After They're Built*, Viking, 1994.
4. The R-value (resistance to heat) of a wall built with standard American domestic wooden construction practices averages 19, while three-string straw bale construction faced on both sides with plaster and cob has been rated with R-values from 52 to 74. See Nehemiah Stone, "Thermal Performance of Straw Bale Wall Systems," *Ecological Building Network* (October 2003), ecobuildnetwork.org.
5. Coppicing is the traditional practice of harvesting multiple generations of

wood from the same tree root system, and enables a relatively small woodlot to produce very large volumes of wood sustainably for centuries. There are few old crafts that deserve a quicker revival.

6. For the New Alchemists see Nancy Jack Todd, ed., *The Book of the New Alchemists*, Dutton, 1977; John Todd and Nancy Jack Todd, *Tomorrow is our Permanent Address*, HarperCollins, 1980. For Earthships see Michael Reynolds, *Earthship: How to Build Your Own, Volume 1*, Solar Survival Architecture, 1990.
7. James Howard Kunstler, *The Geography of Nowhere: The Rise and Decline of America's Man-Made Landscape*, Simon and Schuster, 1993.
8. See especially Leckie et al. 1975 and Olkowski et al. 1979.
9. It is of course possible to get "raspberry" jam much more cheaply by buying commercial products whose ingredients consist mostly of corn syrup and artificial flavors. If corn syrup and artificial flavors could be bought directly from the producers and used by home canners, though, similar economies would likely apply.
10. For the following paragraphs I have drawn extensively on Bryan Ward-Perkins, *The Fall of Rome: And the End of Civilization*, Oxford University Press, 2005.

Chapter Eight: Work

1. Energy Information Administration, *Annual Energy Review*, Energy Information Administration, 2007.
2. Daniel Bell, *The Coming of Post-Industrial Society: A Venture in Social Forecasting*, Basic Books, 1973.
3. See, for example, Mary Carruthers and Jan M. Ziolkowski, eds., *The Medieval Craft of Memory: An Anthology of Texts and Pictures*, 2002; and Frances Yates, *The Art of Memory*, Routledge and Kegan Paul, 1966.
4. The American Radio Relay League (ARRL) website atarrl.org is the best online introduction to amateur radio.

Chapter Nine: Energy

1. For German technology see Ian V. Hogg, *German Secret Weapons of the Second World War: The Missiles, Rockets, Weapons and New Technology of the Third Reich*, Greenhill, 2002. For German CTL methods see Peter W. Becker, "The role of synthetic fuel in World War II Germany," *Air University Review*, July–August 1981 (online edition, airpower.maxwell.af.mil).
2. See, for example, the discussion in Charles B. MacDonald, *The Battle of the Bulge*, Phoenix, 1998.
3. Net energy analyses on German CTL technology during the war have ap-

parently not been done, so this is simply a guess; the actual figures would be fascinating to know.

4. These figures are from David Hughes, "Coal: Some inconvenient truths" (Presentation to the Association for the Study of Peak Oil and Gas-USA fourth annual conference, Sacramento, CA, 22 September 2008).
5. Gene Tyner, Sr., "Net energy analysis of nuclear and wind systems," Ph.D. dissertation, University of Oklahoma, 1985.
6. Hosein Shapouri et al., *Estimating the Net Energy Balance of Corn Ethanol*, US Department of Agriculture, 1995.
7. The best discussion remains David Roe, *Dynamos and Virgins*, Random House, 1984.

Chapter Ten: Community

1. See villageforum.com for this plan.
2. Rob Hopkins, *The Transition Handbook: From Oil Dependency to Local Resilience*, Chelsea Green, 2008.
3. Korten, *The Great Turning*.

Chapter Eleven: Culture

1. Korten, *The Great Turning*, uses the "developmental stage" argument.
2. Dijkstra, *Idols of Perversity*, p. 146.
3. In Arthur Koestler, *The Roots of Coincidence: An Excursion into Parapsychology*, Random House, 1972.
4. This passage is preserved in the tenth-century Codex Victorianus Laurentianus, and has been recorded by Gregorio Paniagua and the Atrium Musicale de Madrid in their *Musique de la Grèce Antique* (Arles, France; Harmonia Mundi, 1979).
5. One of the few discussions of this massive problem is Adrian Atkinson, "Collapse and the fate of cities," *City* 12:1 (April 2008), pp. 79–106.
6. See, in particular, John Lukacs *Historical Consciousness: The Remembered Past*, Harper & Row, 1968.

Chapter Twelve: Science

1. James Lovelock "A book for all seasons," *Science* 280 (1998), pp. 832–833.
2. Greer, *The Long Descent* pp. 182–187.
3. See Lewis Mumford, *The Myth of the Machine Volume One: Technics and Human Development*, 1967 and *The Myth of the Machine Volume Two: The Pentagon of Power*, 1972, Harcourt, Brace and World; and Theodore Roszak, *Where the Wasteland Ends: Politics and Transcendence in Post-Industrial Society*, Doubleday, 1972.

4. See, for example, Derrick Jensen and George Draffan, *Welcome to the Machine*, Chelsea Green, 2004.
5. A standard history is William Kneale and Martha Kneale, *The Development of Logic*, Oxford University Press, 1962.
6. D.T. Mason, "Appropriate ecology: a modest stress-strain proposal." *Bulletin of Marine Science* 31 (1981), pp. 723–729.
7. James McClenon, *Deviant Science: The Case of Parapsychology*, University of Pennsylvania Press, 1984, offers a useful case study of the split between inquiry and ideology in modern scientific research.
8. See Roger Penrose, *The Emperor's New Mind: Concerning Computers, Minds, and the Laws of Physics*, Oxford University Press, 1989, for a thoughtful discussion.
9. Thomas Kuhn, *The Structure of Scientific Revolutions*, University of Chicago Press, 1962.
10. McClenon, *Deviant Science*.

Chapter Thirteen: The Ecotechnic Promise

1. See, for example, Timothy Burns, ed., *After History? Francis Fukuyama and his Critics*, Rowman and Littlefield, 1994.
2. "We are history's actors...when we act, we create our own reality." This embarrassing display of hubris by a Bush administration staffer is quoted in Ron Suskind, "Faith, certainty, and the presidency of George W. Bush," *New York Times Magazine* (17 October 2004).
3. Kurzweil, *The Singularity is Near*, is representative.
4. See particularly Karl Popper, *The Poverty of Historicism*, Basic Books, 1957.
5. For a representative sample see Nicholas Campion, *The Great Year: Astrology, Millenarianism, and History in the Western Tradition*, Penguin, 1994.
6. Polybius' anacyclosis may be read in Book 6 of his *The Histories*; see Polybius, *The Histories*, tr. W.R. Paton, Heinemann, 1922–1927.
7. Giambattista Vico, *New Science*, Penguin, 1999, p. xi.
8. Reprinted in Vico, op. cit., p. xxxvi.
9. See especially Gregory Bateson, *Mind and Nature: A Necessary Unity*, Bantam, 1979; Gregory Bateson and Mary Catherine Bateson, *Angels Fear: Towards an Epistemology of the Sacred*, Macmillan, 1987; and Ervin Laszlo, *An Introduction to Systems Philosophy*, Gordon and Breach, 1972.

Bibliography

Atkinson, Adrian, "Collapse and the fate of cities," *City* 12:1 (April 2008), pp. 79–106.

Bateson, Gregory, *Mind and Nature: A Necessary Unity*, Bantam, 1979.

——— and Mary Catherine Bateson, *Angels Fear: Towards an Epistemology of the Sacred*, Macmillan, 1987.

Becker, Peter W., "The role of synthetic fuel in World War II Germany," *Air University Review*, July-August 1981 (online edition, airpower.maxwell.af.mil).

Beckford, James A., *New Religious Movements and Rapid Social Change*, SAGE Publications, 1986.

Bell, Daniel, *The Coming of Post-Industrial Society: A Venture in Social Forecasting*, Basic Books, 1973.

Bell, Graham, *The Permaculture Garden*, Thorsons, 1994.

Brand, Stewart, *How Buildings Learn: What Happens After They're Built*, Viking, 1994.

——— ed., *The Next Whole Earth Catalog*, Rand McNally, 1980.

Brierley, Corale L., et al., *Coal: Research and Development to Support National Energies Policy*, National Academies Press, 2007.

Brown, Lester, *Plan B: Rescuing a Planet Under Stress and a Civilization in Trouble*, Norton, 2003.

Burns, Timothy, *After History? Francis Fukuyama and his Critics*, Rowman and Littlefield, 1994.

Callenbach, Ernest, *Ecotopia*, Banyan Tree, 1975.

Calvin, William H., *A Brain for All Seasons: Human Evolution and Abrupt Climate Change*, University of Chicago Press, 2002.

Campbell, Stu, *Let It Rot!: The Gardener's Guide to Compost*, Storey Publishing, 1990.

——— *The Mulch Book: A Complete Guide for Gardeners*, Storey Publishing, 1991.

Campion, Nicholas, *The Great Year: Astrology, Millenarianism, and History in the Western Tradition*, Penguin, 1994.

Carroll, R. L., *Vertebrate Paleontology and Evolution*, W. H. Freeman and Co., 1988.

Carruthers, Mary, and Jan M. Ziolkowski, eds., *The Medieval Craft of Memory: An Anthology of Texts and Pictures*, University of Pennsylvania Press, 2002.

Catton, William R., Jr., *Overshoot: The Ecological Basis of Evolutionary Change*, University of Illinois Press, 1982.

Costanza, R., d'Arge, R., de Groot, R., Farber, S., Grasso, M., Hannon, B., Limburg, K., Naeem, S., O'Neill, R., Paruelo, J., Raskin, R., Sutton, P., and van den Belt, M., "The value of the world's ecosystem services and natural capital," *Nature* 387 (1997), pp. 253–260.

Darley, Julian, *High Noon for Natural Gas: The New Energy Crisis*, Chelsea Green, 2004.

Deffeyes, Kenneth, *Hubbert's Peak: The Impending World Oil Shortage*, Princeton University Press, 2003.

deMoll, Lane, ed., *Rainbook: Resources for Appropriate Technology*, Schocken, 1977.

de Santillana, Giorgio, and Hertha von Dechend, *Hamlet's Mill: An Essay Investigating the Origins of Human Knowledge and Its Transmission Through Myth*, David Godine, 1977.

Diamond, Jared, *Collapse: How Societies Choose to Fail or Succeed*, Penguin, 2005.

Dijkstra, Bram, *Idols of Perversity: Fantasies of Feminine Evil in Fin-de-Siècle Culture*, Oxford University Press, 1986.

Duhon, David, *One Circle: How to Grow a Complete Diet in Less Than 1,000 Square Feet*, Ecology Action, 1985.

Duncan, Richard C., "The life-expectancy of industrial civilization: the decline to global equilibrium," *Population and Environment* 14:4 (1993), pp. 325–357.

Echenberg, Myron J., *Plague Ports: The Global Urban Impact of Bubonic Plague, 1894–1901*, New York University Press, 2007.

Economic Research Service, United States Department of Agriculture, "Briefing Room: Organic farming and marketing," ers.usda.gov/Briefing/Organic

Energy Information Administration, *Annual Energy Review*, Energy Information Administration, 2007.

Fichman, Martin, *Evolutionary Theory and Victorian Culture*, Humanity Books, 2002.

Friedman, Thomas, *The World Is Flat: A Brief History of the Twenty-first Century*, Farrar, Straus and Giroux, 2005.

Fukuyama, Francis, "The End of History?" *The National Interest*, Summer 1989.

——— *The End of History and the Last Man*, Free Press, 1992.

Galbraith, John Kenneth, *The Great Crash 1929*, Houghton Mifflin, 1954.

Gerson, Mark, ed., *The Essential Neoconservative Reader*, Perseus, 1997.
Gill, Richardson B., *The Great Maya Droughts: Water, Life, and Death*, University of New Mexico Press, 2000.
Gould, Stephen Jay, *Time's Arrow, Time's Cycle: Myth and Metaphor in the Discovery of Geological Time*, Harvard University Press, 1987.
——— *Wonderful Life: The Burgess Shale and the Nature of History*, W. W. Norton, 1990.
Greer, John Michael, *The Long Descent: A User's Guide to the End of the Industrial Age*, New Society Publishers, 2008.
Hayakawa, S. I., *Language in Thought and Action*, Harcourt, Brace, and World, 1964.
Heilbrun, Jacob, *They Knew They Were Right: The Rise of the Neocons*, Doubleday, 2008.
Heinberg, Richard, *The Party's Over: Oil, War, and the Fate of Industrial Society* New Society Publishers, 2003.
——— *Peak Everything: Waking Up to the Century of Declines*, New Society Publishers, 2007.
Hirsch, Robert L., Roger Bezdek, and Robert Wendling, *Peaking of World Oil Production: Impacts, Mitigation, and Risk Management*, US Department of Energy, 2005.
Hogg, Ian V., *German Secret Weapons of the Second World War: The Missiles, Rockets, Weapons and New Technology of the Third Reich*, Greenhill, 2002.
Hopkins, Rob, *The Transition Handbook: From Oil Dependency to Local Resilience*, Chelsea Green, 2008.
Hornborg, Alf, *The Power of the Machine: Global Inequalities of Economy, Technology, and Environment*, Alta Mira Press, 2001.
Hughes, David, "Coal: Some Inconvenient Truths," presentation to the Association for the Study of Peak Oil and Gas-USA fourth annual conference, September 22, 2008.
Hutchins, Robert Maynard, ed., *Machiavelli and Hobbes*, Encyclopedia Britannica, 1952.
Ibn Khaldûn, *The Muqaddimah: An Introduction to History*, tr. Franz Rosenthal, Princeton University Press, 1967.
Intergovernmental Panel on Climate Change, *Climate Change 2007*, Cambridge University Press, 2007.
Ito, Akihiko, and Takehisa Oikawa, "Global mapping of terrestrial primary productivity and light-use efficiency with a process-based model," in M. Shiyomi et al., eds., *Global Environmental Change in the Ocean and on Land*, Terrapub, 2004.
Jeavons, John, *How To Grow More Vegetables Than You Ever Thought Possible on Less Land Than You Can Imagine*, Ten Speed, 1979.

Jenkins, Joseph, *The Humanure Handbook: A Guide to Composting Human Manure*, Chelsea Green, 1999.

Jevons, William Stanley, *The Coal Question: An Inquiry Concerning the Progress of the Nation and the Probable Exhaustion of Our Coal-Mines*, Macmillan, 1866.

Johnson, Warren, *Muddling Toward Frugality*, Shambhala, 1978.

Jung, C. G., "Wotan," in C. G. Jung, *Civilization in Transition*, Princeton University Press, 1970, pp. 179–193.

Kelly, John, *The Great Mortality: An Intimate History of the Black Death, the Most Devastating Plague of All Time*, HarperCollins, 2005.

Keyhoe, Alice Beck, *The Ghost Dance: Ethnohistory and Revitalization*, Holt, Rinehart and Winston, 1989.

King, F. H., *Farmers of Forty Centuries: Organic Farming in China, Korea, and Japan*, Rodale Press, 1973.

Kneale, William, and Martha Kneale, *The Development of Logic*, Oxford University Press, 1962.

Koestler, Arthur, *The Roots of Coincidence: An Excursion into Parapsychology*, Random House, 1972.

Korten, David, *The Great Turning*, Berrett-Koehler, 2006.

Kotschar, V. F., *Fighting with Property: A Study of Kwakiutl Potlatching and Warfare, 1792–1930*, University of Washington Press, 1950.

Kuhn, Thomas, *The Structure of Scientific Revolutions*, University of Chicago Press, 1962.

Kunstler, James Howard, *The Geography of Nowhere: The Rise and Decline of America's Man-Made Landscape*, Simon and Schuster, 1993.

Kurzweil, Ray, *The Singularity is Near*, Viking, 2005.

Laszlo, Ervin, *An Introduction to Systems Philosophy*, Gordon and Breach, 1972.

Le Conte, Joseph, *Evolution*, D. Appleton and Co., 1888.

Le Corbusier, *Towards a New Architecture*, Dover, 1986.

Leckie, Jim, Gil Masters, Harry Whitehouse and Lily Young, *Other Homes and Garbage: Designs for Self-Sufficient Living*, Sierra Club Books, 1975.

Lewis, C. S., *The Abolition of Man*, Macmillan, 1947.

——— *That Hideous Strength*, Scribner, 1996.

Lovelock, James, "A book for all seasons," *Science* 280 (1998), pp. 832–833.

Lukacs, John, *Historical Consciousness: The Remembered Past*, Harper & Row, 1968.

MacDonald, Charles, B., *The Battle of the Bulge*, Phoenix, 1998.

Mason, D. T., "Appropriate ecology: a modest stress-strain proposal," *Bulletin of Marine Science* 31 (1981), pp. 723–729.

McClenon, James, *Deviant Science*, University of Pennsylvania Press, 1984.

Meadows, Donnella, David Meadows, Jorgen Randers and William W. Behrens III, *The Limits to Growth*, Universe, 1972.

Mills, C. Wright, *The Causes of World War Three*, Simon and Schuster, 1958.
Mollison, Bill, *Permaculture Two*, International Tree Crops Institute, 1979.
——— and David Holmgren, *Permaculture One*, Transworld, 1978.
Mumford, Lewis, *The Myth of the Machine Volume One: Technics and Human Development*, Harcourt, Brace and World, 1967.
——— *The Myth of the Machine Volume Two: The Pentagon of Power*, Harcourt, Brace and World, 1972.
Neiberg, Michael S. and Dennis B. Showalter, *Soldiers' Lives Through History: The Nineteenth Century*, Greenwood, 2006.
Odum, Eugene P., *Fundamentals of Ecology*, W.B. Saunders, 1971.
Odum, Howard, *Environmental Accounting: Emergy and Environmental Decision Making*, John Wiley and Sons, 1996.
Van der Ryn, Sim, Helga Olkowski, Bill Olkowski, Tom Javits and Farallones Institute, *The Integral Urban House: Self Reliant Living in the City*, New Society Publishers, 2008.
Orlov, Dmitry, *Reinventing Collapse: The Soviet Example and American Prospects*, New Society Publishers, 2008.
Ortega y Gasset, José, *The Modern Theme*, Harper, 1961.
Orwell, George, "Politics and the English Language," *Horizon*, April, 1946.
Parks, Alexia, *Rapid Evolution: Seven Words That Will Change Your Life Forever*, Education Exchange, 2002.
Penrose, Roger, *The Emperor's New Mind: Concerning Computers, Minds, and the Laws of Physics*, Oxford University Press, 1989.
Piekalkewicz, Jaroslaw and Alfred Wayne Penn, *The Politics of Ideocracy*, State University of New York Press, 1995.
Plato, *Collected Dialogues*, ed. Edith Hamilton and Huntington Cairns, Princeton University Press, 1961.
Polybius, *The Histories*, tr. W.R. Paton, Heinemann, 1922–1927.
Ponting, Clive, *A Green History of the World: The Environment and the Collapse of Great Civilizations*, St. Martin's, 1992.
Popper, Karl, *The Poverty of Historicism*, Basic Books, 1957.
Quinn, Daniel, *Ishmael*, Bantam, 1992.
Ray, Paul H. and Sherry Ruth Anderson, *Cultural Creatives: How 50 Million People Are Changing the World*, Harmony Books, 2000.
Reynolds, Michael, *Earthship: How to Build Your Own, Volume 1*, Solar Survival Architecture, 1990.
Roe, David, *Dynamos and Virgins*, Random House, 1984.
Roszak, Theodore, *Where the Wasteland Ends: Politics and Transcendence in Post-Industrial Society*, Doubleday, 1972.
Schilthuis, Willy, *Biodynamic Agriculture*, Floris, 1994.
Seymour, John, *The Self-Sufficient Gardener*, Doubleday, 1978.

Shapouri, Hosein, James A. Duffield and Michael S. Grabowski, *Estimating the Net Energy Balance of Corn Ethanol*, US Department of Agriculture, 1995.

Simmons, Nancy B., "Taking wing: Uncovering the evolutionary origins of bats," *Scientific American*, December, 2008 (online edition, sciam.com).

Spengler, Oswald, *The Decline of the West*, tr. Charles Francis Atkinson, Knopf, 1926–1929.

Steiner, Rudolf, *Agriculture: An Introductory Reader*, Biodynamic Agriculture Association, 1958.

Steinfeld, Carol, *Liquid Gold: The Lore and Logic of Using Urine to Grow Plants*, Ecowaters, 2007.

Stone, Nehemiah, "Thermal Performance of Straw Bale Wall Systems," *Ecological Building Network*, (October 2003), ecobuildnetwork.org.

Strachan, David, "Hay fever, hygiene, and household size," *British Medical Journal* 299 (1989), pp. 1259–1260.

Suskind, Ron, "Faith, certainty, and the presidency of George W. Bush," *New York Times Magazine*, 17 October, 2004.

Tainter, Joseph A., *The Collapse of Complex Societies*, Cambridge University Press, 1988.

Taleb, Nassim Nicholas, *The Black Swan: The Impact of the Highly Improbable*, Random House, 2007.

Taylor, Graeme, *Evolution's Edge: The Coming Collapse and Transformation of Our World*, New Society Publishers, 2008.

Telleen, Maurice, *The Draft Horse Primer*, Rodale Press, 1977.

Thornburg, Newton, *Valhalla*, Little, Brown, 1980.

Todd, John, and Nancy Jack, *Tomorrow is our Permanent Address: The Search for an Ecological Science of Design as Embodied in the Bioshelter*, HarperCollins, 1980.

Todd, Nancy Jack, ed., *The Book of the New Alchemists*, Dutton, 1977.

Toynbee, Arnold J., *A Study of History*, 10 vols., Oxford University Press, 1934–55.

Tuchman, Barbara W., *The Proud Tower: A Portrait of the World Before the War, 1890–1914*, Macmillan, 1962.

Tudge, Colin, *Neanderthals, Bandits, and Farmers: The Origins of Agriculture*, Yale University Press, 1998.

Tyner, Gene, Sr., "Net energy analysis of nuclear and wind systems," Ph.D. dissertation, University of Oklahoma, 1985.

Vacca, Roberto, *The Coming Dark Age*, tr. J. S. Whale, Doubleday, 1973.

Vico, Giambattista, *New Science*, Penguin, 1999.

von Franz, Marie-Louise, "The Process of Individuation," in C. G. Jung, ed., *Man and his Symbols*, Doubleday, 1964.

Wallace, Anthony F. C., *Revitalization Movements*, Bobbs-Merrill, 1956.

Ward-Perkins, Bryan, *The Fall of Rome: And the End of Civilization*, Oxford University Press, 2005.
Whitefield, Patrick, *Permaculture in a Nutshell*, Permanent Publications, 1997.
Wright, Frank Lloyd, *Truth Against the World: Frank Lloyd Wright Speaks for an Organic Architecture*, Wiley-Interscience, 1987.
Yates, Frances, *The Art of Memory*, Routledge and Kegan Paul, 1966.
Ziarek, Ewa Plonowska, *An Ethics of Dissensus: Postmodernity, Feminism, and the Politics of Radical Democracy*, Stanford University Press, 2001.

Index

F

G

H

I

M

N

O

P

R

S

T

U

V

W

Y

Z

Printed in the USA
CPSIA information can be obtained
at www.ICGtesting.com
JSHW011914221024
72178JS00004B/38